NEXT-GENE] SEQUENCING AND SEQUENCE DATA ANALYSIS

Authored By

Kuo Ping Chiu

Associate Research Fellow/Associate Professor
Genomics Research Center, Academia Sinica
National Taiwan University
National Central University
Taiwan

including, without limitation, special, incidental and/or consequential damages and/or damages for lost data and/or profits arising out of (whether directly or indirectly) the use or inability to use the Work. The entire liability of Bentham Science Publishers shall be limited to the amount actually paid by you for the Work.

General:

1. Any dispute or claim arising out of or in connection with this License Agreement or the Work (including non-contractual disputes or claims) will be governed by and construed in accordance with the laws of the U.A.E. as applied in the Emirate of Dubai. Each party agrees that the courts of the Emirate of Dubai shall have exclusive jurisdiction to settle any dispute or claim arising out of or in connection with this License Agreement or the Work (including non-contractual disputes or claims).

2. Your rights under this License Agreement will automatically terminate without notice and without the need for a court order if at any point you breach any terms of this License Agreement. In no event will any delay or failure by Bentham Science Publishers in enforcing your compliance with this License Agreement constitute a waiver of any of its rights.

3. You acknowledge that you have read this License Agreement, and agree to be bound by its terms and conditions. To the extent that any other terms and conditions presented on any website of Bentham Science Publishers conflict with, or are inconsistent with, the terms and conditions set out in this License Agreement, you acknowledge that the terms and conditions set out in this License Agreement shall prevail.

Bentham Science Publishers Ltd.
Executive Suite Y - 2
PO Box 7917, Saif Zone
Sharjah, U.A.E.
Email: subscriptions@benthamscience.org

BENTHAM SCIENCE

CONTENTS

ABOUT THE AUTHOR

 **Kuo Ping Chiu** is currently an Associate Research Fellow at Academia Sinica with joint appointments with National Taiwan University (NTU) and National Central University (NCU). He received his PhD in Microbiology from UC Davis in 1991 and completed his postdoc at Harvard Medical School on Neurosciences during 1993 - 1996. Kuo Ping's research career is closely associated with biotechnology. His PhD research focused on intracellular amplification of mouse mammary tumor proviral DNA using *in situ* PCR to identify the infected cells, while his postdoctoral training was related to multiple colorimetric labeling of acetylcholine receptor subunit transcripts to study their coordinated expression pattern. These trainings have critical influence on his academic research career and industrial experience. His industrial experience started with a job at Bio-Rad Laboratories where he developed protocols and kits for antimicrobial susceptibility testing using flow cytometry (5/1996-4/1998). Later he switched from wetlab to Bioinformatics and worked for Genome Institute of Singapore on developing Paired-End diTag technology and methods for sequence data analysis (8/2002 - 8/2008). He moved back to Taiwan in 2008 working for Academia Sinica on developing DNA sequencing-related biotechnologies and studying gene expression and regulation in normal and cancer cells. He also teaches sequencing technologies, sequence data analysis and pathway analysis in a number of national universities including NTU, NCU, and National Yang-Ming University (NYMU). He is currently holding three US patents related to paired-end ditag technologies.

FOREWORD

I have known Kuo Ping for almost thirty years. He is very fond of science, especially in developing or employing novel technologies for biological investigations. He and I often share ideas and have worked on common scientific interests, including the project to study the early anti-cancer activity of the medicinal fungus *Antrodia cinnamomea* using Next-Generation Sequencing (NGS) to sequence and compare the transcriptomes of cancer cells treated with or without the extract of *A. cinnamomea*. This work led to the discovery that this medicinal fungus was able to globally collapse the miRNA system.

It is his never-resting mind that has provided him a motivation to start writing this book some years ago, while he was still working for Genome Institute of Singapore when NGS was starting to take shape. During that time, there was nowhere to find a book to introduce advanced sequencing technologies and the potential applications to either students or laymen. This book evolves from his previous work on paired-end ditag technology and has a strong association with NGS which can be considered as the most fascinating technology of the 21st century. I am glad to see him accomplish this hard task. This book seems to be a good textbook for graduate students and a handbook for researchers who are interested in sequencing and biotechnologies.

It is worth mentioning that Kuo Ping is a scientist with a strong passion for outdoor activities. We did some interesting, and sometimes funny, things together. Back in California, we spent some weekends fishing at Pacifica Pier for perch and crabs, and then barbecued and talked over beer in my backyard. Here in Taiwan, we often tour around his "secret gardens" where he grew trees and raised domestic fish in a pond. The most interesting outdoor experience was a trip to find fossil stones in a wild countryside, where we found forgotten creeks running through the hillside. We tied a very heavy fossil stone with rope and carried it with a bamboo stick across water streams, making loud laughter in the wild. I actually believe that his passion for outdoor activities has contributed to the completion of this useful book.

Miao-Lin (Merlin) Hu, Ph.D.
Honorary Distinguished Professor,
National Chung Hsing University
Department of Food Science and Biotechnology
250 Kuo Kuang Road, Taichung 402
Taiwan

PREFACE

Many biological fields, including genetics, immunology, genomics, and epigenomics, can be readily integrated with bioinformatics and sequencing. In fact, sequencing is becoming an indispensable tool for biological and medical investigations, especially for the study of cancer, genetics, epigenetics, immunology, and developmental biology. Foreseeing the potential applications of sequencing technologies, we have initiated a sequence data analysis course at the Institute of Zoology, National Taiwan University to introduce sequencing technologies, genomics, epigenomics, bioinformatics, and biotechnologies to graduate and undergraduate students. At the same time, I started organizing the teaching materials to compose this book. Most of the materials were derived from my research at Academia Sinica and my teaching at National Taiwan University (NTU), National Central University (NCU), and National Yang Ming University (NYMU). Additionally, there are some from my previous work in the United States and Singapore.

Besides college students, this book is also written for people who already have some biological background, or with an interest in knowing more about DNA sequencing and its applications. With this effort, we hope to help develop sequence-associated knowledge among people from all walks of life, including researchers, professionals, and amateurs working in biology-related or -unrelated fields, with or without an ambition to apply sequencing technologies to biological or medical investigations. Extracting biological meaning from large quantities of sequence data is an art. We intend to help readers appreciate the various types of sequencing technologies, and learn how to use sequencing technologies to unravel the mystery of the biological system.

Sequence data analysis is not only an art, but also a tool to help researchers develop a constructive philosophy. When a diploma is awarded to a candidate, it may be treated by the recipient as a degree which can then help the graduate to find a better job, or achieve a higher social status. Some recipients may take it more seriously and use it for further, and possibly more advanced, studies. Similarly, sequence data analysis can be treated as either a course to be accomplished in class, or, alternatively, a process which will be a guide to the further understanding of the life of nanoscopic molecules within a cell. We gain many more experiences through the years of working in a lab, and, with those proficiencies, we build our philosophy. Sequencing and sequence data analysis can help us make sense of the activities which take place on a daily basis in the molecular world of our bodies, even though these activities are impalpable to the naked eye. Being able to "see" things that cannot "normally" be seen can help us build a philosophy, in which we travel between the macroscopic and molecular worlds.

A sincere attitude towards sequence data analysis is essential. Since omics sequence datasets are normally at large volumes, these data cannot be handled by traditional means, but by computer programs. Frequently we find bugs in computer programs which may produce erroneous or misleading results. A bug is a bug. No matter whether it is big or small, it has to be removed, so as to obtain the real molecular status of interest. Moreover, the author would like to emphasize the importance of practical exercise. Students are strongly encouraged to personally construct sequencing libraries, run sequencing, and analyze sequence libraries or related libraries whenever possible.

Although this book carries a mission, readers can treat it either as a novel, or a science fiction story. The development of sequencing technologies per se is science fiction, isn't it? With limitations in man power and time, the contents of this edition may not be able to completely satisfy serious readers. Your opinions and suggestions are warmly welcomed!

Kuo Ping Chiu, Ph.D.
Genomics Research Center, Academia Sinica
National Taiwan University
National Central University
Taiwan

ACKNOWLEDGEMENTS

I would like to thank Grace Chu-Fang Lo, former Dean of the College of Life Science (CLS) of NTU (National Taiwan University), and Director Pan from the Institute Zoology of NTU for their support on running a course on sequence data analysis at NTU. Although most people have realized the importance of building high throughput sequencing facilities in Taiwan, financial supports remain very limited. Professor Lo was among the few scientists who can fully appreciate the effort to build a NGS sequencing facility at NTU campus. I also like to express my special thanks to Hong Sain Ooi and Prof. Xiaodong Zhao, my longtime friends since we worked together back in Singapore, for their support on RNA-Seq analysis and friendship. Academia Sinica (AS) president CH Wong, AS Vice President CJ Chen, AS Genomics Research Center (GRC) Vice Director Alice LT Yu and Prof. ML Hu from National Chung Hsing University also provided great help on my career at Academia Sinica. Prof. H.T. Yu from the Institute of Zoology of NTU, CLS secretary Ho, my friend Amy Y.M. Chou from biotech industry, Dean Pei from National Pingtung University of Science and Technology and many of my colleagues at AS and NTU have also been very kind and helpful. Thank all of you!

CONFLICT OF INTEREST

The author confirms that this ebook contents have no conflict of interest.

Part 1

Background Introduction

The Ultimate Frontier of Knowledge: the Mysterious Genomes and the Gene Expression and Regulation in the Upstream of Biological Information Flow

Abstract: DNA sequencing is just a tool that we can employ to study biological phenomena. It can be useful for us to review some biological background before we discuss how to use the tool. It will help us to understand what subjects in the biological field DNA sequencing can be applied to and how to apply sequencing technologies in the study. Before we discuss the applications of DNA sequencing, Let's review some molecular biology, starting from the structure and organization of information molecules at the molecular level.

Keywords: Alternative splicing, Central dogma, Chromatin, DNA packaging, Epigenetic modifications, Euchromatin, Gene, Genome, Heterochromatin, MicroRNA, Nucleosome, Transcription factors.

Definition of Terminologies

Genome

A genome is a complete set of genetic material in a cell. Prokaryotic cells do not have nuclei or mitochondria, while eukaryotic cells have both. The genome of an E. coli cell is a circular chromosome, while a eukaryotic cell normally comprises of a set of linear nuclear chromosomes together with a circular mitochondrial chromosome (chrM).

Chromatin

A chromatin is an interphase chromosome, comprising DNA and proteins bound to the DNA. Sometimes chromatin and chromosome are interchangeably used in the book.

Central Dogma

The classic view of gene expression from DNA to RNA, and then to protein.

Gene

In a broad sense, a gene is a genomic region capable of being transcribed into mRNA (protein-coding messenger RNA) or non-coding RNA. In other words, a gene is a piece of molecular information and a transcription unit, prepared and stored in the genome for being transcribed into mRNA, which has a protein-coding capability, or a non-coding RNA, which potentially has regulatory or structural function. According to gene nomenclature, gene names should be in lower case and italicized, but sometimes the first letter can be capitalized. On the other hand, protein names should have the first letter or the whole name capitalized and should not be italicized.

Transcript
The product of transcription.

Non-Coding RNA
Synonymous to non-protein-coding RNA, a non-coding RNA (ncRNA) is a RNA transcript which is not destined for translation. In contrast, a protein-coding RNA is a message molecule capable of being translated into a protein. ncRNAs include rRNA, tRNA, miRNA (microRNA), siRNA, snRNA, piRNA, etc.

Locus (plural: loci)
Simply means a genetic location.

Homolog
A gene (or protein) homolog is a nucleotide (or amino acid) sequence related to another nucleotide (or amino acid) sequence by at least partial sequence homology.

Ortholog
An ortholog is a common ancestor-derived transcription unit found in multiple species. Normally they have the same, or similar, functions.

Paralog
Gene paralogs are normally generated by gene duplication. Due to the redundancy in structure and function, gene paralogs may diversify through evolution.

Gene Isoforms
Gene isoforms are transcription units encoding proteins with similar functions and thus can be assigned to the same position of a pathway.

Scales Used in the Molecular World
Angstrom (10^{-10} meter, about the diameter of a hydrogen atom), nano-meter (10^{-9} m, or ~10 H atoms), micro-meter (10^{-6} m), *etc.*

INTRODUCTION

Transcription, or gene expression, represents the forefront of biological information flow. Information flow at this stage is carried out predominantly by chromatin-associated activities inside the nucleus. The mRNA molecules are subsequently translocated from the nucleus to the cytoplasm and get translated by ribosomes into proteins. Gene expression in the upstream is regulated in a hierarchical manner through the interaction between various biological molecules, especially RNAs and proteins. Conceivably, information molecules, such as DNA and RNA, are appropriate for DNA sequencing when studying gene expression and regulation.

MICROSCOPIC STRUCTURE AND ORGANIZATION OF INFORMATION MOLECULES

Chromatins are packed into distinguishable domains with variable degrees of condensation. Each of these domains, either heterochromatins or euchromatins, may occupy a large portion of the chromatin. Moreover, some domains may switch from heterochromatin to euchromatin, or *vice versa* , and are thus defined as facultative heterochromatin. Heterochromatin domains are condensed and more likely to be located underneath the nuclear envelope (Fig. 1), although some are dispersed across the nuclear matrix. On the other hand, euchromatin domains are loosely packed and are most likely to be found in the interior of the nucleus or near the nuclear pores.

Genes within the heterochromatin domains are transcriptionally inactive, while those in the euchromatin domains are accessible for regulatory proteins and transcription machinery. It is conceivable that genes encompassed in the heterochromatin domains are correlated with tissue type as determined during development and differentiation. Moreover, both the content of heterochromatin-encompassed and the euchromatin-encompassed genes can be dynamically influenced by intracellular and extracellular signals. Induced pluripotent stem cells (iPS) are a good example.

When compared to heterochromatin, euchromatin is gene-rich, more open to transcription machinery, while heterochromatin is condensed chromatin only visible under a microscope. Heterochromatin represents an inactive state of interphase chromatin and exists in every centromere and the inactivated X-chromosome. Some euchromatic regions become heterochromatic in later life (implicated in development and aging). Heterchromatin contains very few genes (most in the facultative heterochromatins) and favors HP1 binding. Heterochromatins can be either facultative (reversible, *e.g.* X inactivation) or constitutive (fixed and irreversible). Inactivation of one of the X chromosome is mediated by epigenetic modification. The inactivated X chromosome varies from one cell to another. Thus, a female "tissue" is a mosaic construct containing a mixture of gene products from both X chromosomes. Activated genes move from the peripheral nuclear region to the interior region, or from the facultative heterochromatin region to the euchromatin region and require chromatin remodelers, all containing multiple subunits, for activation (Narlikar *et al.* 2002).

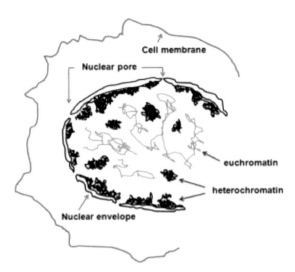

Fig. (1). Distribution of heterochromatins and euchromatins in the nucleus. This schematic drawing demonstrates the locations of heterochromatin and euchromatin domains in the nucleus. Most heterochromatin domains are underneath the nuclear envelope, with some dispersed in the nuclear matrix, while most of the euchromatin domains are loosely dispersed in the nuclear matrix.

THE STRUCTURE AND ORGANIZATION OF INFORMATION MOLECULES AT THE MOLECULAR LEVEL

DNA is the Most Magnificent Molecule Selected By Evolution

As selected by evolution for the storage of biological information, DNA is probably the most beautiful molecule in the world (Fig. **2**).

In response to the environmental conditions, DNA molecules may switch between A, B, and Z form (Table **1**). Under normal physiological conditions, DNA molecules are in B form, having 10 bp per turn and with 3.4 angstrom (0.34 nm) thickness between base pairs oriented approximately perpendicular to its helical axis. The major groove and minor groove, both running in parallel in right-handed conformation along the DNA molecule, are of continuous helical structure if not interrupted by A- or Z-form segment. When water is less available, DNA conformation shifts from B to A form, which remains right-handed, but the thickness of the base pairs shrink from 3.4 angstroms to 2.7 angstroms, and the angle between base pairs and the sugar-phosphate backbone tilts about 30°. Under heavy methylation, DNA may shift from B to Z form, which is left-handed and the thickness of base pairs becomes 3.8 angstroms. Given the fact that the alpha helix secondary structure of a protein is also in right-handed conformation and thus is expected to best fit the B form structure, a conformational switch between A, B, and Z forms also plays a role in regulating gene expression. For more detailed information, please refer to Molecular Biology of the Gene by James D. Watson *et al.*, and many other great publications.

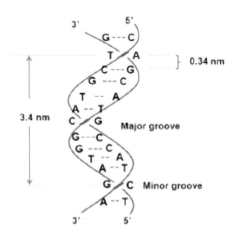

Fig. (2). DNA double helix.

Table 1. DNA conformation.

Form	Right or left handed?	Condition	bp/ turn	Base pair distance (angstrom)
B	Right-handed	under normal physiological conditions	10	3.4
A	Right-handed	under dehydrated conditions	11	2.7
Z	Left-handed	Under high ionic strength	12	3.8

Genes live in the genome. Similar to families living in a valley, genes can be found in both strands of the chromosome (just like families living on both sides of a valley) and gene densities in complementary strands are more or less correlated (just like the formation of villages along the floor of the valley), *i.e.*, if a region in one strand (say, the plus strand) is gene-rich, then the same region of the other strand (the minus strand) is very likely to be gene-rich as well. Notice that, the position of a nucleotide in a chromosome, no matter it is in the plus or minus strand, is based on its location in the "plus" strand, from left (5' end) to right (3' end), counted base by base.

DNA Packaging

Here, DNA packaging refers to the packaging of chromatin (Fig. **3**). Evolution has developed a mechanism to pack chromosomes in a structurally and functionally efficient manner within the nucleus. Given the fact that the thickness of a B-form DNA is about 3.4 angstrom (1 angstrom = 1×10^{-10} m) per base pair and that a diploid human genome contains about 6×10^9 base pairs ($2 \times 3 \times 10^9$ bp), the total length of an unpacked human genome in the nucleus of a single cell is about two meters, equivalent to about 10 centimeters per chromosome. After packaging into a mitotic chromosome, the length shrinks to just a few micrometers. Notice that, eukaryotic *cells* are typically a *few tens of microns* in size. Without packaging, chromosomes would become entangled with one another and won't be able to fit into the nucleus.

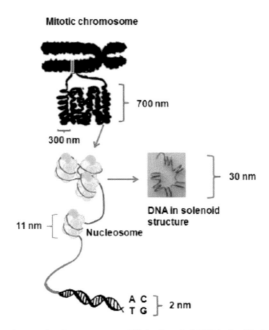

Fig. (3). Packaging of a eukaryotic chromosome. Right-handed DNA double helix first forms nucleosome structure by wrapping ~1.7 times around histone octamer in left-handed supercoil conformation. Nucleosomes further coiled into 30 nm solenoid chromatin fiber (~6 nucleosomes per turn), which in turn loops into 300 nm scaffolds and then chromosome.

Even during interphase, chromatins have to be packed in some way to fit well into the nucleus while simultaneously maintain functional efficiency and inter-chromatin coordination. In fact, chromatin architecture is very dynamic and responds to the molecular environment. Even transcriptionally active regions may still be encompassed within 80-100 nm chromonema fibers (Horn and Peterson, 2002). Genomic DNAs are bound by histone proteins which are implicated in DNA-folding and (about equal amounts of) non-histone proteins such as DNA polymerases or RNA polymerases. Nucleosomes are the basic building blocks of chromatin. DNAs are negatively charged, while most DNA-binding proteins, including histone proteins H1, H2A, H2B, H3, and H4, are positively charged - due to K/R-rich. Fitting a total of 1 - 2 meters of DNA into the nucleus requires extensive DNA folding which is primarily mediated by the interaction of negative and positive charges.

Nucleosome

A nucleosome is formed by wrapping 146-147 bp DNA in left-handed supercoil conformation around a histone core (two copies of each H2A, H2B, H3, and H4) about 1.7 times (Fig. 4), achieving a compacting factor of ~5 (147 x 3.4 Å / 100 Å per octamer).

Fig. (4). Schematic presentation of the nucleosome structure and histone N-terminal " tails".

For the formation of a nucleosome, the $(H3/H4)_2$ tetramer forms first and combines with $(H2A/H2B)_2$ tetramer and then H1. Variants are commonly found in histone proteins including H1 (*e.g.* H5) and H3 (*e.g.* H3.3 and Cid). Wrapping of double helix DNA on histone octamer is left-handed and is negative supercoiling (unwinding). The 30 nm (300 Å) filament is a further left-handed helix for the string of nucleosomes, with H1 histone proteins sitting in the center and act as glue to secure the chromatin fiber structure. The formation of ≥ 300 Å filaments, which are commonly found in euchromatins, represents an inactive state. It is believed that, left-handed DNA supercoiling on a histone core, or an even higher level supercoiling, is a method to store negative-supercoiling (W) energy for unwinding DNA, which is essential for replication, transcription, and DNA repair. The **L**inking number is the sum of the **T**wists (the number of turns around the central axis of the double helix) and the **W**rithing number (the number of turns of the **S**upercoil structure). The storage of energy requires topoisomerase II (or simply called gyrase), which consumes ATP to generate negative supercoils.

Nucleosome positioning is partially determined by the genomic sequence (Segal *et al.*, 2006). Studies have further indicated that epigenetic modifications (EMs) and transcription factors play the key a role in the dynamic modulation of DNA architecture. EM is carried out by specific enzymes. Some transcription factors (TFs) have methyltransferase or acetyltransferase activity and may directly modify the tails of histone proteins in the nucleosome, resulting in a switch in the packaging pattern. As such, these TFs do not directly work on DNA, but on histones instead. Through the modifications they made on the histone proteins, which in turn affect the "packaging density", these TFs influence the transcription potential of their target genes.

GENE EXPRESSION AND REGULATION IN THE EUKARYOTIC SYSTEM

The biological system is extremely complex. It is not possible, and in fact not necessary, to go through every detail. Here, let's review some key points instead.

A. ATP-Dependent and ATP-Independent Chromatin Remodeling Enzyme

An ATP-dependent chromatin remodeling enzyme can cause non-covalent modification on a chromatin structure and localized fine tuning of a chromatin structure. It is able to displace nucleosomes and causes "sliding" of DNA with respect to the histone octamer, creating a conformation where DNA is accessible for transcription factors. Examples of ATP-dependent chromatin remodelers include: 1) yeast SWI (yeast mating type switching)/ SNF (sucrose non-fermenting) complex (Smith *et al.*, 2003), or BAF in human. 2) RSC (Remodels Structure of Chromatin) (Asturias *et al.*, 2002; Skiniotis *et al.*, 2007). SWI/SNF complex is recruited to the promoter by DNA-bound activators or repressors (Martens and Winston, 2003). A bromodomain present in the SWI/SNF ATPase subunit is needed for stable binding of SWI/SNF to the acetylated lysine (K) in histones. The bromodomain is also present in TFs (Martens and Winston, 2003). In Drosophila, expression of a dominant-negative mutant of BRM, SWI2/SNF2 protein in Drosophila, severely hindered the association of RNA pol II with chromatin, suggesting the presence of a sequential order for the formation of transcription complex.

On the other hand, an ATP-independent chromatin remodeling enzyme causes covalent modifications. It can result in a large-scale alteration in the chromatin structure. Most of this work focuses on histones, especially on H3 and H4. Examples include: 1) HAT (histone acetyltransferase) and 2) HDAC (histone deacetylase).

B. X Chromosome Inactivation

X chromosome inactivation is an interesting phenomenon in eukaryotic biology. It is a cis-limited silencing of an entire (X) chromosome that may affect more than 1000 genes to achieve a monoallelic X-linked gene expression for dosage compensation between males and females. In humans, the inactive X chromosome varies from one cell to another, so to create a "mosaic" cellular phenotype. Xist non-coding RNA expressed in one of the two X chromosome (random in human) and localized to the X-chromosome inactivation center (Xic) and spread out in both directions, resulting in the exclusion of euchromatic histone marks, including H3 and H4 acetylation (H3K9ac), H3K4me2, and H3K4me3, and simultaneous replacement by other types of histone modifications such as H3K27me3, H3K9me2, ubiquitination of H2A, and H4K20me1. Methylation of CpG islands (in the 5' ends of genes) affects the DNA structure (*e.g.* transition from B-form to other forms). In mice, Tsix (antisense of Xist) represses Xist and prevents the active X chromosome from being inactivated. In human females, they may be co-expressed and interact differently.

C. Pathway Flow from Extracellular Signal to the Nucleus

For both single-celled and multi-cellular organisms, pathways are frequently initiated by ligand binding to their corresponding receptors on the cell membrane. Ligand binding induces receptor dimerization or oligomerization and causes a relay of the signal to cytosolic signaling molecules, leading to the activation and nuclear translocation of specific transcription factors. Transcription factors then bind to their target motifs and recruit epigenetic modifiers, cofactors, and RNA polymerase II to initiate the transcription of a set of responsive, or positively regulated, genes and the repression of another set of negatively regulated genes. Pathway activation and repression can be analyzed by plugging transcriptome (gene expression) data into a pathway database.

D. TFs, Motifs, and DNA-Protein Interaction

Most of the interactions between DNA and transcription factors occur in the major groove. As all natural amino acids are in L-form , protein alpha helices are right-handed. It has been a common phenomenon that right-handed protein helices inserts into the right-handed major groove, or, less frequently, minor groove, of DNA structure where specific motifs are located. Most motifs comprise short palindromic sequences to interact with specific TF dimers. Some DNA motifs do not have a distinguishable palindromic structure. Under such circumstances, multiple proteins in the complex may participate in DNA binding. Motifs, with *de novo* or known, can be analyzed with a ChIP-TFBS approach. How does evolution select the components of a transcriptional complex? It is known that selection of a certain transcription start site (TSS) and polyadenylation site (PAS) responds to different extracellular signals. Notice that, beyond the interactions between DNA and DNA-binding proteins, transcription factories impose another layer of transcriptional control.

E. Transcription Factories (TFs)

Transcription, with a speed of ~20-50 millisecond per base, or about "½ - 1 step (region covered by a transcription machinery, ~ 100 bp) per second", is highly regulated and has to occur in the right place at the right time. First of all, each chromatin/chromosome occupies a relative territory (Dundr and Misteli, 2001) and transcription factories are dynamically formed (or disappeared) for the transcription (or silencing) of specific genes (Cook, 1999). In particular, nucleoli are special transcription factories that specialized for 1) making rRNAs, 2) processing rRNAs, and possibly, 3) the initiation of assembling processed rRNAs with ribosomal proteins. Activities in transcription factories require transcription factors and the binding sites of a specific transcription factor can be analyzed with ChIP-TFBS libraries.

F. Epigenetic Modifications

Each chromosome can be divided into dispersed active or inactive domains (genomic regions) by epigenetic modifications (EM), which includes histone modifications (HM) and DNA methylation (DM) to form a control device that turns genes on or off over a long range of chromosome without affecting the genetic sequence. Both HM and DM regulate multiple biological processes including transcription, replication, development, differentiation, and apoptosis. Both EM and DM can be analyzed with ChIP-mediated approaches.

HMs are mediated by histone tails which are subjected to a number of covalent modifications such as methylation, acetylation, ubiquitination, ADP ribosylation, phosphorylation, SUMOylation, ADP ribosylation, de-amination, and proline isomerization (Table 2). Most of these histone modifications can be either repressive or active, except acetylation, which always results in transcriptional activation. HMs in CA1 neurons of the hippocampus are implicated in the storage of long-term memory.

Table 2. Epigenetic modifications.

Type of modification	Target residues	Enzymes	Effects
DNA methylation	C (to 5meC) in CpN (N=A, T, or G) of CpG islands	DNA methyltransferases (reversed by DNA demethylases)	Most likely to be negative
Acetylation on histone	K in H3, H4	Acetyltransferase	Positive only
Methylation on histone	K/R in H3, H4	Methyltransferases	Can be either positive or negative
Phosphorylation on histone	S/T in H3, H4	Kinases	Can be either positive or negative
Ubiquitination on histone	K (in H2A, H2B, H3 and H1)	Ubiquitinase E2,3	Can be either positive or negative
Sumoylation on histone	K	SUMO-conjugating and deconjugating enzymes	Can be either positive or negative
ADP ribosylation on histone	E	ADP-ribosyltransferase	Can be either positive or negative
Deamination on histone	R	Deaminases	Can be either positive or negative
Proline isomerization	P	Prolyl isomerases	Can be either positive or negative

Ref: (Arnaudo and Garcia, 2013; Bottomley, 2004; Kouzarides, 2007).

Methylation in CpG islands normally results in inhibition of gene expression. CpG islands are commonly found in the 5' proximal region of housekeeping genes and genes frequently expressed. In mammals, CpG islands constitute about 1-2% of the genome. Multiple lines of evidence showed that various types of EMs are correlated. For example: 1) DNAme-binding protein was found in a complex that also has histone acetylation activity; 2) H2Bub is associated with H3me at specific K residues (Kinyamu *et al.*, 2005). Thus, cascades of EMs seem to exist, but remain unclear.

In summary, for coordinated regulation, certain well-defined chromatin regions are dynamically opened and closed, while some regions are permanently "locked" for certain types of tissues or cells. 'Binary switches' and 'modification cassettes' mechanisms explain why the EM variegation is selected (Fischle *et al.*, 2003).

G. Alternative Splicing

Spliceosome is composed of snRNAs (including U1, U2, U4, U5, and U6) and proteins. During transcription, spliceosome is in close proximity to the transcription machinery. Genome-wide transcriptome analysis revealed a high level of mRNA complexity caused by alternative splicing in metazoan (Kalsotra and Cooper, 2011). Alternative splicing variants can be readily identified by a paired-end ditag approach during transcriptome analysis. Similarly, alternative splicing variants can be identified by cufflinks software.

H. MicroRNA (miRNA)

Mature miRNAs are small single-stranded non-coding RNAs of ~18–24 nucleotides known to posttranscriptionally regulate up to 50% of genes in both plants and animals. MicroRNA exerts posttranscriptional repression by binding to the 3' UTR of its target mRNA followed by direct degradation or translational inhibition of its target mRNA. Many miRNAs have a base variation at one or both ends, forming various types of isomirs. Similar to proteincoding genes, miRNA biosynthesis is mediated by RNA polymerase II (Pol II) which transcribes miRNA genes to generate primary miRNAs (pri-miRNAs). Immature miRNAs form a stem loop structure. This unique feature has been employed to distinguish miRNA sequences from non-specific sequences. MicroRNAs can be analyzed with miRNA libraries

Maturation of miRNA transcripts first occur in the nucleus and continue through their subsequent stay in the cytoplasm. In the nucleus, complex of Drosha and DGCR/Pasha cleaves the pri-RNA to produce 70 nt hairpin-shaped precursor miRNAs (premiRNAs), which are then transported by Exporin-5 to the cytoplasm, where the pre-miRNAs are cleaved by the complex of Dicer and TRBL/Loquacious, releasing the double stranded, 21 bp (miRNA-miRNA* duplex) mature miRNAs. In most cases, the miRNA* strand is degraded, whereas the 59 end of single-stranded mature miRNA is incorporated into RNA-induced silencing complex (RISC) with Argonaut protein to regulate its target mRNA. Through binding to the 3'UTR of its target gene, a miRNA can either degrade the target mRNA or repress its translation.

Compared to mRNAs, the turnover rate of miRNAs is, in general, much faster. XRN2 was first reported as an essential factor which specifically degrades single-stranded mature miRNAs in Caenorhabditis elegans. miRNAs degrades by XRN2 and can modulate functional miRNAs homeostasis in animals. Moreover, XRN2 targeting those miRNAs failed to load into Argonaute and unbound to their target genes for degradation.

In our study of the anti-cancer effects of *Antrodia cinnamomea* fungus, we found that the miRNA system can be globally collapsed by the ingredient compounds of *A. cinnamomea* which interact with multiple miRNA metabolic pathways (Chen *et al.*, 2013).

THE BOTTOM LINE

Again, this chapter aims to review some biological background so that readers can better understand how to apply DNA sequencing to biological studies. Transcription of specific DNA regions of a genome into RNA molecules represents the upmost stage of gene expression. All RNA molecules can be converted into DNA sequences, thus all transcriptomes can be sequenced and analyzed. Additionally, by using an antibody as a mediator to pull out DNA-binding proteins (the target of that antibody) and thus the DNA segments associated with the target proteins, followed by sequencing the associated ChIP fragments, one can identify many types of DNA segments in the genome. This approach has become a powerful method for the study of regulatory mechanisms which control gene expression. Thus, sequencing is able to resolve DNA and RNA sequences as well as the regulatory mechanisms in a quantum leap fashion and its resolution can reach to the nucleotide level.

REFERENCES

Arnaudo, A.M., Garcia, B.A. (2013). Proteomic characterization of novel histone post-translational modifications. *Epigenetics Chromatin,* *6*(1), 24.
[http://dx.doi.org/10.1186/1756-8935-6-24] [PMID: 23916056]

Asturias, F.J., Chung, W.H., Kornberg, R.D., Lorch, Y. (2002). Structural analysis of the RSC chromatin-remodeling complex. *Proc. Natl. Acad. Sci. USA,* *99*(21), 13477-13480.
[http://dx.doi.org/10.1073/pnas.162504299] [PMID: 12368485]

Bottomley, M.J. (2004). Structures of protein domains that create or recognize histone modifications. *EMBO Rep.,* *5*(5), 464-469.
[http://dx.doi.org/10.1038/sj.embor.7400146] [PMID: 15184976]

Chen, Y.J., Thang, M.W., Chan, Y.T., Huang, Y.F., Ma, N., Yu, A.L., Wu, C.Y., Hu, M.L., Chiu, K.P. (2013). Global assessment of Antrodia cinnamomea-induced microRNA alterations in hepatocarcinoma cells. *PLoS One,* *8*(12), e82751.
[http://dx.doi.org/10.1371/journal.pone.0082751] [PMID: 24358224]

Cook, P.R. (1999). The organization of replication and transcription. *Science,* *284*(5421), 1790-1795.
[http://dx.doi.org/10.1126/science.284.5421.1790] [PMID: 10364545]

Dundr, M., Misteli, T. (2001). Functional architecture in the cell nucleus. *Biochem. J.,* *356*(Pt 2), 297-310.
[http://dx.doi.org/10.1042/0264-6021:3560297] [PMID: 11368755]

Fischle, W., Wang, Y., Allis, C.D. (2003). Binary switches and modification cassettes in histone biology and beyond. *Nature,* *425*(6957), 475-479.
[http://dx.doi.org/10.1038/nature02017] [PMID: 14523437]

Horn, P.J., Peterson, C.L. (2002). Molecular biology. Chromatin higher order folding--wrapping up transcription. *Science,   297*(5588), 1824-1827.
[http://dx.doi.org/10.1126/science.1074200] [PMID: 12228709]

Kalsotra, A., Cooper, T.A. (2011). Functional consequences of developmentally regulated alternative splicing. *Nat. Rev. Genet.,   12*(10), 715-729.
[http://dx.doi.org/10.1038/nrg3052] [PMID: 21921927]

Kinyamu, H.K., Chen, J., Archer, T.K. (2005). Linking the ubiquitin-proteasome pathway to chromatin remodeling/modification by nuclear receptors. *J. Mol. Endocrinol.,   34*(2), 281-297.
[http://dx.doi.org/10.1677/jme.1.01680] [PMID: 15821097]

Kouzarides, T. (2007). Chromatin modifications and their function. *Cell,   128*(4), 693-705.
[http://dx.doi.org/10.1016/j.cell.2007.02.005] [PMID: 17320507]

Martens, J.A., Winston, F. (2003). Recent advances in understanding chromatin remodeling by Swi/Snf complexes. *Curr. Opin. Genet. Dev.,   13*(2), 136-142.
[http://dx.doi.org/10.1016/S0959-437X(03)00022-4] [PMID: 12672490]

Narlikar, G.J., Fan, H.Y., Kingston, R.E. (2002). Cooperation between complexes that regulate chromatin structure and transcription. *Cell,   108*(4), 475-487.
[http://dx.doi.org/10.1016/S0092-8674(02)00654-2] [PMID: 11909519]

Segal, E., Fondufe-Mittendorf, Y., Chen, L., Thåström, A., Field, Y., Moore, I.K., Wang, J.P., Widom, J. (2006). A genomic code for nucleosome positioning. *Nature,   442*(7104), 772-778.
[http://dx.doi.org/10.1038/nature04979] [PMID: 16862119]

Skiniotis, G., Moazed, D., Walz, T. (2007). Acetylated histone tail peptides induce structural rearrangements in the RSC chromatin remodeling complex. *J. Biol. Chem.,   282*(29), 20804-20808.
[http://dx.doi.org/10.1074/jbc.C700081200] [PMID: 17535815]

Smith, C.L., Horowitz-Scherer, R., Flanagan, J.F., Woodcock, C.L., Peterson, C.L. (2003). Structural analysis of the yeast SWI/SNF chromatin remodeling complex. *Nat. Struct. Biol.,   10*(2), 141-145.
[http://dx.doi.org/10.1038/nsb888] [PMID: 12524530]

CHAPTER 2

Evolution of DNA Sequencing Technologies

Abstract: Current nucleotide sequencing focuses on DNA sequencing. As you will see later in this chapter, direct RNA sequencing was de-selected by evolution. Instead, RNA molecules are converted to cDNA and subject to DNA sequencing. Moreover, DNA sequencing can be conducted by a number of sequencing technologies, each of which uses a company- or inventer-defined procedure and sequencing mechanism. By nature, both living organisms and non-living objects are constantly challenged by evolution. DNA sequencing technologies are of no exception. For living organisms, phenotypic variations resulted from genetic alterations are constantly tested by the surrounding environment, which allows the fittest to propagate more efficiently than the others. Similarly, using DNA sequencing technologies that we will discuss later in this chapter as an example, each technology has its pros and cons against one another. Also, their advantages and disadvantages co-evolve with, and depend on their environmental backgrounds. Here, we review the evolution of DNA sequencing technologies to appreciate the evolutionary process eventually leading to the development of Next-Generation Sequencing technologies.

Keywords: Next-generation sequencing, NGS, Sanger sequencing, Single-molecule sequencing.

Definition of Terminologies

Sequencing

Sequencing, in biological terms, refers to the usage of methodologies and/or instruments to determine the order of building blocks in a macromolecule. DNA sequencing determines the order of deoxyribonucleic acid bases (A, T, G, and C) in a DNA molecule, RNA sequencing determines the order of ribonucleic acid bases (A, U, G, and C) in a RNA molecule, and protein sequencing determines the order of amino acids in a protein molecule. The authors will focus on DNA sequencing using the Next-Generation Sequencing (NGS) technologies.

Sanger Sequencing

In short, Sanger sequencing can be defined as the method that uses di-deoxynucleotides (ddNTPs) in partial termination reactions. This approach is readily distinguishable from the NGS sequencing methods which use engineered substrates instead of ddNTPs. Sanger sequencing can be further categorized into manual Sanger sequencing (using radioactive labeling) and automated Sanger sequencing (using four-colored fluorescent labeling in

Kuo Ping Chiu

conjunction with computerized signal-capturing and processing system).

INTRODUCTION

As one may be aware, the application of sequencing technologies is enormous. Genome sequencing has unraveled the genetic sequences of hundreds of prokaryotic and eukaryotic organisms, and many more are on the way. Genome assemblies are used as references for further biological and medical investigations. So far, a number of prokaryotic and eukaryotic genomes have been sequenced. With the momentum provided by NGS technologies, the number of sequenced genomes is increasing dramatically. Sequencing of specific molecular species produced along the gene expression and regulation cascade play an important role in unraveling the entities of the molecular species and the study of vertical and horizontal interactions between molecules. Undoubtedly, these efforts will strengthen our understanding of certain key subjects in the "biological field", including variation in immune repertoire, genetic diversity, developmental process, and diseases.

CHARACTERISTICS OF DNA SEQUENCING

There are evidently clear characteristics for DNA sequencing. These include, but not limited to, 1) completeness, 2) high resolution, and run in quantum jump fashion. By running DNA sequencing base-by-base, we can completely read through a genome. Moreover, DNA sequencing is able to provide a resolution at, and beyond, the single nucleotide level. For example, DNA sequencing allows access to the locations of SNVs across the whole genome. Furthermore, modifications on nucleotide (*e.g.*, DNA methylation) or amino acids (*e.g.*, methylation, acetylation, phosphorylation, *etc*) can be analyzed. In contrast to hybridization, DNA sequencers perform sequencing in quantum jump fashion.

ADVANCES IN SEQUENCING TECHNOLOGIES

In fact, there was RNA sequencing before the commencement of DNA sequencing. RNA sequencing adopted the following procedure: 1) Label RNA sample (*e.g.*, bacterial phage) with radioactivity, (*e.g.*, P^{32}). 2) Treat sample with chemicals and ribonucleases to hydrolyze RNA at specific residues. 3) Run the sample with 2D gel/membrane. 4) Use the amino acid sequence of the corresponding protein as a reference to help understand the RNA sequence. Overall, the procedure is tedious, making it not a competitive methodology (Metzker, 2010).

DNA sequencing prevailed over RNA and protein sequencing due to, at least, the following reasons:

1. All amino acids (about 20 or so) are encoded only by 4 deoxynucleotides through 3-

base coding. This 4-to-20 encoding mechanism makes it easy to convert a nucleotide/deoxynucleotide sequence to its corresponding amino acid sequence, but not the other way around. As such, deoxynucleotides/nucleotide sequencing makes more sense.

2. Technically, DNA sequencing is much simpler than RNA or protein sequencing. DNA polymerase was the only enzyme required for DNA sequencing and, compared to RNA and protein sequencing, preparation of sequencing reagents and setup of reaction conditions are much easier for DNA sequencing.

3. DNA is much more stable than RNA, and as such, is much easier than dealing with RNA.

4. RNA can be easily converted to cDNA (complementary DNA).

5. In general, DNA molecules, either double-stranded or single-stranded, are structurally simpler than single-stranded RNA molecules, which are likely to loop back and form complex structures. Notice that base sequencing requires the target molecules (either in DNA or RNA form) to be single-stranded; double-stranded structures are expected to interfere the sequencing process.

6. Polymerase chain reaction (PCR), which amplifies DNA molecules, dramatically facilitates DNA sequencing.

MANUAL SANGER SEQUENCING

In nature, DNA replication is conducted by DNA polymerase which uses deoxyribonucleotides triphosphates (dNTP, namely dATP, dTTP, dGTP, and dCTP) as building blocks for the synthesis of the complementary strand. This process requires the replication origins (ori) to be opened by helicase and the synthesis be primed (*i.e.* started from specific sites) by "primers". The synthesized strand can elongate only in 5' to 3' direction.

Following the publication of DNA double helix structure in 1953 by Francis Crick and James Watson, Frederick Sanger (Fig. 1) invented an enzymatic method for DNA sequencing in 1975 - 1977 (Sanger and Coulson, 1975; Sanger *et al.*, 1977). He took advantage of this natural system and invented a method for DNA sequencing which helped him to win a Nobel Prize. During the same period of time Maxam and Gilbert invented a chemical method for DNA sequencing (Maxam and Gilbert, 1977), which was de-selected by competition because of its technical complexity.

For the purpose of DNA sequencing, Sanger made a number of modifications. The procedure can be summarized as the following steps:

1. A single species of DNA sample was prepared in single-stranded form, which would be used as templates for the synthesis of the second DNA strand.

2. A sequence-specific oligonucleotide (oligo) was prepared to serve as the 'sequencing primer'. This primer flanks the upstream (5') of the DNA region to be sequenced. Such

sequencing primer defines the sequencing starting site for all four reactions (see below).

3. The primer was labeled with isotope to distinguish the synthesized strand from the template.

4. Di-deoxynucleotide triphosphates (ddNTPs, namely, ddATP, ddTTP, ddGTP, and ddCTP) were added separately to make four reactions (ddATP + dNTP), (ddTTP + dNTP), (ddGTP + dNTP), and (ddCTP + dNTP). In each reaction, an accidental incorporation of a di-deoxynucleotide triphosphate by DNA polymerase would result in termination of the sequencing, because the 3'-OH group was not available for further elongation. This phenomenon is called "partial termination". Partial termination makes DNA synthesis of some molecules to be terminated, while the others in the same tube continue. Since each termination represents a specific location in the DNA, no fragments are of the same size.

5. After the sequencing reaction, double-stranded, double helix DNA molecules (template plus the newly synthesized strand) were denatured so to release the radioactive newly synthesized strand for gel electrophoresis.

6. Four reactions were displayed separately by polyacrylamide gel electrophoresis. Since the mobility is reversely proportional to fragment size, shorter fragments were expected to be found in the bottom of the gel.

7. The gel was dried by heat under vacuum.

8. The dried gel was exposed on an X-ray film so the radioactive signal could be caught by the X-ray film.

9. Bands in the X-ray film were read from bottom up, four lanes intercalated as one.

Fig. (1). Frederick Sanger.

AUTOMATED SANGER SEQUENCING

The original/manual Sanger sequencing could only sequence a short stretch of DNA molecule of about a few hundred bases in length, and each run normally took at least a full week. Conceivably, automation is essential for high throughput DNA sequencing (Fig. **2**). The first semi-automated sequencing was made possible by Leroy Hood and colleagues in the California Institute of Technology (CIT) (Smith *et al.*, 1986). In 1998,

Prizm 377 gel sequencer & 3700 capillary array sequencer were manufactured by ABI (Applied Biosystems). These sequencers facilitated the progress of the human genome sequencing and shortened the Human Genome Project (HGP) from 10 years to just a few years. It all relied on sequencing automation (Mardis, 2013)!

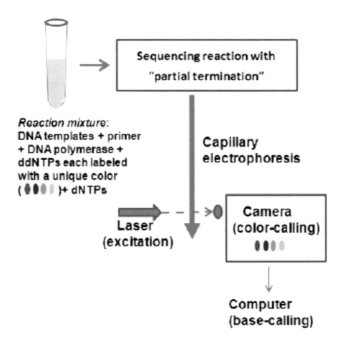

Fig. (2). Sequencing automation.

KEY CHANGES FROM MANUAL TO AUTOMATED SANGER SEQUE-NCING

A few changes were made from manual to automated Sanger sequencing:

1. Four distinct fluorescent colors were used separately to label the ddNTPs.
2. Four reactions can now be combined together in a single tube.
3. A capillary gel, instead of a polyacrylamide gel, was used to display the single-stranded fragments.
4. A laser beam was setup near the end of the capillary gel to excite the incorporated fluorescent ddNTPs.
5. Emissions of fluorescent signals were captured by a computerized camera.
6. The computer stored colorimetric signals and converted the "color codes" into nucleotide sequences.

THE NEXT-GENERATION SEQUENCING

Next-Generation Sequencing (NGS, or Next-Gen Sequencing) technologies are now making a greater impact on sequencing. The invention of the 454 sequencer in 2005 by

Jonathan Rothberg represents a hallmark in modern sequencing technology. In fact, NGS is made of a number of robust technologies and, in general, is characterized by high speed and high yield. However, both the sequencing cost and error rate remain high but acceptable. Since all NGS sequencers can generate large amounts of sequence data in a short period of time, read number is no longer a concern. The issue is how to improve bioinformatics tools to facilitate genome assembly and the understanding of gene expression and regulation.

Initially, NGS platforms adopt two types of sequencing mechanisms: by synthesis or by ligation (Fig. 3). The former is employed by 454 and Solexa systems, while the latter by SOLiD system. Details of NGS will be described in the next chapter (Huang *et al.*, 2012).

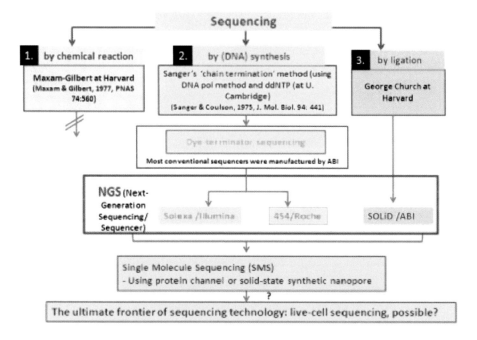

Fig. (3). Development of sequencing technologies.

Although NGS has been a powerful technology, all NGS sequencers still require PCR to amplify DNA molecules to overcome the limitation in the strength of fluorescent signal and the sensitivity of fluorescence detection. Moreover, current error rate for NGS sequencing is still staying more or less around 1% and thus require high yield to cover this deficiency (see next chapter).

SINGLE MOLECULE SEQUENCING

Single Molecule Sequencing (SMS), or more precisely 'Single DNA Molecule Sequencing (SDMS)', refers to the usage of a sequencing technology to decipher the nucleotide sequence of a single DNA molecule without PCR amplification. SMS technologies will not only unleash DNA sequencing from the requirement of the PCR amplification, they are also expected to produce much longer reads (a few thousand bases per read) with much lower cost and dramatically reduced error rate.

Single molecule sequencing can be achieved by various ways, as a number of companies and research laboratories have been working around the clock to develop and improve multiple SMS technologies over a number of years.

Among these SMS approaches is the nanopore DNA sequencing. A nanopore simply refers to a pore with an inner diameter at the nanometer scale. It can be made from either synthetic material or a membrane protein (*e.g.* using alpha hemolysin (αHL) or Mycobacterium smegmatis porin A (MspA)). The cross diameter of a DNA double helix is about two nanometers (nm), or 2×10^{-9} m, and that of a single-stranded DNA molecule is even less, both small enough to pass through a nanopore for the purpose of sequencing. As expected, DNA is more suitable for nanopore sequencing because RNA, as it is normally single-stranded but with high potential of forming secondary structures which would interfere with the sequencing. Moreover, DNA is more stable than RNA.

It is expected that some years later when these nanotechnologies become more sophisticated, we should be able to use single molecule sequencing not only for longer reading at lower cost, but also for practically filling gaps left in the genome assemblies by NGS sequencing.

REFERENCES

Huang, Y.F., Chen, S.C., Chiang, Y.S., Chen, T.H., Chiu, K.P. (2012). Palindromic sequence impedes sequencing-by-ligation mechanism. *BMC Syst. Biol., 6* (Suppl. 2), S10.
[http://dx.doi.org/10.1186/1752-0509-6-S2-S10] [PMID: 23281822]

Mardis, E.R. (2013). Next-generation sequencing platforms. *Annu. Rev. Anal. Chem. (Palo Alto, Calif.), 6*, 287-303.
[http://dx.doi.org/10.1146/annurev-anchem-062012-092628] [PMID: 23560931]

Maxam, A.M., Gilbert, W. (1977). A new method for sequencing DNA. *Proc. Natl. Acad. Sci. USA, 74*(2), 560-564.
[http://dx.doi.org/10.1073/pnas.74.2.560] [PMID: 265521]

Metzker, M.L. (2010). Sequencing technologies - the next generation. *Nat. Rev. Genet., 11*(1), 31-46.
[http://dx.doi.org/10.1038/nrg2626] [PMID: 19997069]

Sanger, F., Coulson, A.R. (1975). A rapid method for determining sequences in DNA by primed synthesis with DNA polymerase. *J. Mol. Biol., 94*(3), 441-448.
[http://dx.doi.org/10.1016/0022-2836(75)90213-2] [PMID: 1100841]

Sanger, F., Nicklen, S., Coulson, A.R. (1977). DNA sequencing with chain-terminating inhibitors. *Proc. Natl. Acad. Sci. USA, 74*(12), 5463-5467.

Smith, L.M., Sanders, J.Z., Kaiser, R.J., Hughes, P., Dodd, C., Connell, C.R., Heiner, C., Kent, S.B., Hood, L.E. (1986). Fluorescence detection in automated DNA sequence analysis. *Nature, 321*(6071), 674-679. [http://dx.doi.org/10.1038/321674a0] [PMID: 3713851]

CHAPTER 3

Mechanisms of Next-Generation Sequencing (NGS)

Abstract: DNA sequencing consists of a number of methodologies, each adopts a unique process of sequencing mechanisms. During 1970s, Sanger sequencing survived the competition against other approaches and dominated DNA sequencing for a number of decades. As stimulated by urgent demand of high throughput sequencing approaches by the Human Genome Project and various genome projects that followed, Next-Generation Sequencing evolved to replace Sanger sequencing as the main sequencing approach. NGS consists of three major sequencing platforms (*i.e.*, 454/Roche, Solexa/Illumina and SOLiD/Life Technologies) and each has its own sequencing mechanism. These mechanisms have experienced severe competition, leading to the election of Illumina system by the sequencing market as the main stream sequencing platform. Before we can fully appreciate the reasons leading to the success of the Illumina system, here we analyze and discuss the sequencing mechanisms adopted by these NGS platforms.

Keywords: 454, Bridge amplification, Emulsion PCR, *in situ* PCR, Illumina sequencing, Next-generation sequencing, NGS, Solexa, SOLiD, Sequencing-by-synthesis, Sequencing-by-ligation.

Definition of Terminologies

Target (DNA)
DNA molecules to be sequenced

Sequencing adaptors
Short DNA fragments ligated to the ends of the target DNA. It normally contains both PCR primer-binding sites for PCR amplification and sequencing primer-binding sites for sequencing initiation.

Sequencing library
A library that has been made ready for sequencing. At this point, a pair of sequencing adaptors should have been ligated to the ends of the target molecules.

Sequencing run
A complete sequencing of a (sequencing) library.

Sequencing cycle (or chemistry cycle)
A full process required to complete the incorporation of a nucleotide (for sequencing-by-

synthesis sequencers) or an oligo (for sequencing-by-ligation sequencers) into the elongating strand. (A sequencing run comprises a number of sequencing cycles.)

Massively parallel sequencing
Initially used by 454, to mean a massive number of sequencing reactions simultaneously taking place in massively amplified templates in a synchronized fashion.

INTRODUCTION

Changes Made From Automated Sanger Sequencing to Next-Gen Sequencing

1. No more colony picking
2. Clonal *in situ* PCR amplification of templates becomes essential for all NGS sequencing library preparations. Current NGS sequencing technologies can only sequence DNA templates that have been clonally amplified to thousands of copies in each flow cell channel (Illumina sequencers) or millions of copies on beads (454 and SOLiD machines) by *in situ* PCR, so that signals produced from the massive parallel sequencing reactions in each monoclonal template population can reach beyond the detectable level. Please note that, DNA templates need to be reset to a single-stranded state at two stages: 1) before clonal amplification of templates with *in situ* PCR, and 2) before sequencing.
3. ddNTPs are replaced by ddNTP analogs.
4. Fluorescent signals are still used in Solexa and SOLiD systems, but the 454 system uses light emission as its signal.
5. No electrophoresis is needed. Instead, signals are caught by camera directly on spots.
6. No improvement in sequencing accuracy and the sequence length gets even shorter!
7. However, yield reaches the giga-base level and sequencing cost is dramatically reduced.

In situ PCR for Clonal Amplification of Templates

Strategies of *in situ* PCR for clonal amplification of templates vary among sequencers but can be categorized into two types: 1) solid phase PCR amplification by Solexa system, and 2) emulsion PCR (emPCR or ePCR) by 454 and SOLiD systems (see following sections).

NGS Sequencing Mechanisms

NGS sequencers adopt two types of sequencing mechanisms: by synthesis - using DNA polymerase or by ligation - using DNA ligase (Metzker, 2010). Each category can be further divided into subtypes based on the chemistry undertaken. Since the pros and cons of a NGS sequencer are mainly determined by its sequencing mechanism (Table **1**), it is essential for us to discuss the sequencing mechanisms adopted by each NGS sequencer.

Comparison between different NGS systems can be found in a number of informtive review articles although only a few are listed in the references (Mardis, 2008; Metzker, 2010).

Table 1. Two types of NGS sequencing machines/mechanisms.

Sequencer	454 series	GA/Solexa series	SOLiD series
Inventor/manufacturer	454 Life Sciences -> Roche	Solexa -> Illumina	Applied Biosystems -> Life Technologies
First launch	2005	2006	2007
History	The 454 sequencer was invented by Jonathan Rothberg, the founder of 454 Life Sciences. Dr. Rothberg is the pioneer of NGS machines. Roche acquired 454 Life Sciences in 2007.	Solexa , a spin-off company from Cambridge Univ., launched its first NGS sequencer, Genome Analyzer, in 2006. Solexa was purchased by Illumina in 2007.	SOLiD (supported oligo ligation detection) sequencers are manufactured by Applied Biosystems Inc. (AB or ABI), a longtime leader in DNA sequencing.
Sequencing mechanism	Sequencing by synthesis	Sequencing by synthesis	Sequencing by ligation
Direction of elongation	5' -> 3'	5' -> 3'	3' -> 5'
Direction of reading	5' -> 3'	5' -> 3'	5' -> 3' (on the template strand)
Substrates/ building blocks	dNTPs	dNTP analogs (fluorescently labeled and 3' blocked)	oligos
Delivery of substrates/ building blocks	One by one	All together	All together
Signal	Light (Photon)	Fluorescence	Fluorescence
Imaging	Concurrent w/ the release of signal	Colors taken right after the nt is incorporated	colors taken right after the nt is incorporated
Terminate signal after each cycle?	No	Yes	Yes
One base per cycle?	No	Yes	Yes, if based on sequencing run. No, if based on reaction cycle.
Polymer problem	Yes	No	No
Can read through palindrome?	Yes	Yes	No (major drawback for SOLiD systems!)

I. Sequencing-By-Synthesis

This approach is continuous from the Sanger sequencing method and is adopted by 454 and Solexa sequencers and their derivatives.

<A> Solexa Sequencers Manufactured By Illumina – HiSeq 2000 as an Example (Fig. 1)

Fig. (1). HiSeq 2000.

Early Solexa machines, including Solexa series, Genome Analyzer (GA) series, and HiSeq systems (Fig. **1**) use multi-channeled flow cells (Fig. **2**) for template clonal amplification, or so-called cluster amplification, bridge amplification, or solid phase amplification by the manufacturer. In practice, it is done off-sequencer. Later in MiSeq, clonal amplification is done on the sequencer, making the process simpler and more efficient.

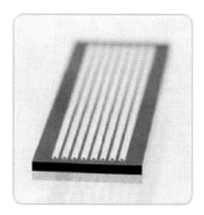

Fig. (2). Illumina GA flow cell.

Sequencing Library Preparation

1. Sonicate target DNA sample if needed.
2. Select DNA fragments of a few hundred bp in length. What size range to choose depends on the max sequencing length of the sequencer. For example, a range between 200-500 bp is suitable for 2013 MiSeq machines.
3. End repair (polish) DNA fragments.
4. Ligate DNA fragments to sequencing adaptors.
5. Load library (containing single-stranded templates) into flow cell channels

Bridge PCR amplification (Fig. 3)

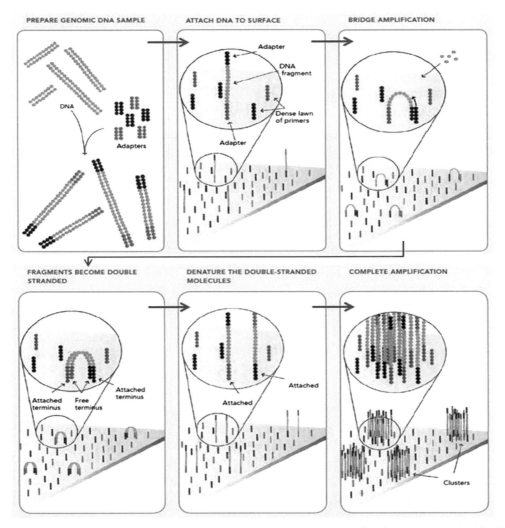

Fig. (3). Cluster generation for Solexa sequencer and derivatives (From Illumina user manual with slight modifications).

i. Oligos with sequences complementary to adaptors are pre-attached on the inner surface of the flow cell. These oligos immobilize the adaptor-ligated template DNA molecules.

ii. PCR amplification creates thousands of copies of target DNA molecules within an area of ~1 micrometer in diameter.

iii. Tens of millions of clusters each originated from a single template are created in the flow cell channel.

iv. Denaturation causes bridges to 'stand up'.

v. An enzymatic cleavage reaction can select a particular strand for sequencing.

vi. Denaturation leaves a selected population of single-stranded DNA molecules covalently attached to the on-surface sequencing adaptors for sequencing.

Notice:

1. *Similar to the 454 system, the Solexa sequences and derivatives do not use ddNTPs. Instead, engineers in their R&D wisely modified the substrates to fulfill two purposes: 1) to let these substrates carry fluorescent signals, and 2) to prevent second polymerization from taking place until the next cycle (not applicable to the 454 machines).*
2. *Unlike the 454 system, there is no homopolymer problem for Illumina's sequencers.*

Sequencing cycles

Note: Prior to sequencing, the camera catches the locations of all clusters. Each cluster is given an ID based on its location relative to the X- and Y-coordinates.

First cycle:

1. Primers bind to single-stranded templates.
2. The presence of all four fluorescently labeled substrates and DNA polymerase is essential. DNA pol incorporates a complementary substrate into the elongating strand.
3. The laser beam excites the fluorescent moiety, resulting in the emission of the fluorescent signal, which is caught by a computerized camera and stored in the computer (Fig. **4**).
4. Locations of fluorescent signals are associated with their corresponding locations and ID (Fig. **4**).
5. Two chemical reactions occur: 1) to trim off the fluorescent signals, and 2) reactivate the 3' terminus of the ribose moiety so that the 3'-OH group becomes available for further elongation (Fig. **5**).

Second cycle and onwards: loop Steps 2-5. (All components are needed except the sequencing primer.)

Note:

1. Termination of the sequencing run: Sequencing will terminate when a pre-determined cycle number is reached.
2. Conversion of color code to nucleotide sequences: sequential color codes produced from sequencing cycles are delineated and converted to bases for all cluster spots.
3. Sequence of the sequencing template can be easily obtained through reverse-complementary conversion process.

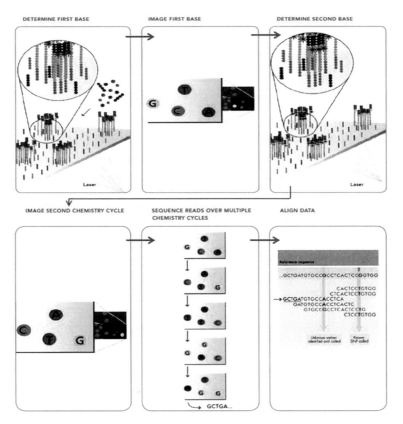

Fig. (4). Sequencing mechanism of Solexa sequencer and derivatives. (From Illumina user manual with slight modifications).

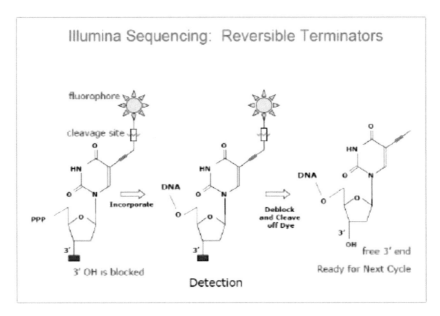

Fig. (5). Fluorescent labeling and 3'-OH block for Illumina/Solexa sequencing. (From http://genetics. stanford.edu/gene211/lectures/Lecture2_HTS_Technologies-2013.pdf).

\<B\> 454 Sequencers By Roche

The 454 system, a sequencing-by-synthesis-based sequencing system invented by Jonathan Rothberg in 2005, can be considered as the first NGS sequencer made commercially available (Ahmadian *et al.*, 2006). This system was later developed into GS FLX (Fig. **6**), which has the highest throughput among 454 sequencers.

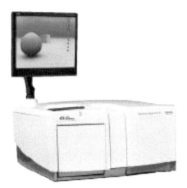

Fig. (6). 454 Genome Sequencer FLX.

Sequencing Library Preparation

1. DNA molecules are fragmented to ~300-500 bp
2. Polish (end repair) DNA fragments
3. Ligate DNA fragments to adaptors A and B. Adaptor B contains 5'-biotin tag to immobilize the adaptor-ligated target DNA molecules onto the streptavidin-coated beads.
4. Run emulsion PCR to amplify ligated DNA on beads (Fig. **7**).
5. Enrich the positive beads, *i.e.* beads with amplified template clusters. Normally clonal amplification can generate millions of copies of target DNA on a single bead.

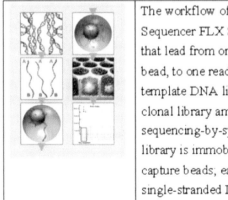

The workflow of 454 Life Sciences' Genome Sequencer FLX System comprises four main steps that lead from one purified DNA fragment, to one bead, to one read. Generation of a single-stranded template DNA library is followed by emulsion-based clonal library amplification, data generation via sequencing-by-synthesis, and data analysis. The DNA library is immobilized on specifically designed DNA capture beads; each bead carries a unique single-stranded DNA library fragment.

Fig. (7). Sequencing mechanism of the 454 system. (From http://www.genengnews.com/gen-articles/third-generation-sequencing-debuts/3257/?page=2).

Sequencing-By-Synthesis on Picotiter Plate

1. Beads with emPCR amplified target DNA are loaded into wells in the picotiter plate, a miniaturized version of a microtiter plate, as the sequencing plate; each picotiter plate contains millions of wells to harbor sequencing beads (> 1 micron in diameter) (Fig. 7). Beads are distributed inside wells in 1:1 ratio
2. Use DNA polymerase to carry out sequencing reaction across all single-stranded template clones (micro-reactors).
3. Four different nucleotides are passed through the picotiter plate one-by-one in sequential order.
4. When an incorporation of a complementary base occurs, PPi is released. As catalyzed by sulfurylase, PPi reacts with APS to generate ATP, which in turn fuels the oxidation of luciferin driven by luciferase. *Notice that, the presence of a homopolymer (e.g. GGG) can trigger the incorporation of multiple bases of the same type (e.g. CCC) in a single reaction. As such, the sequence reads of a library won't be of a uniform size. This is a unique feature of the 454 sequencing.*
5. It is pyrosequencing. 'pyro' means 'fire', because each successful polymerization reaction results in the release of photon(s).
6. Millions of copies of DNA templates in each clone (micro-reactor) are sequenced in a synchronized manner to help elevate the signal to a detectable level.
7. Homopolymers result in variable signal intensity which can be resolved by means of bioinformatics.

II. Sequencing By Ligation: SOLiD Sequencers (Fig. 8) Manufactured By Life Technologies Inc.

Fig. (8). SOLiD 5500xl sequencer.

Compared to sequencing by synthesis, this system is less competitive because of certain intrinsic problems (Huang *et al.*, 2012). Although this system is losing ground and being deselected, its sequencing mechanism remains interesting and deserves discussion as an example to demonstrate the ups and downs of a biotechnology.

i) Sequencing library preparation
    1. Prepare pre-libraries from DNA or RNA samples.
    2. Build a sequencing library (constructs) from each pre-library (*e.g.*, cDNA or ChIP

library)

3. Ligate constructs to beads by ePCR (Fig. **9**). Note: both monoclonal and polyclonal beads are made, but the reaction conditions are optimized to favor monoclonal amplification.

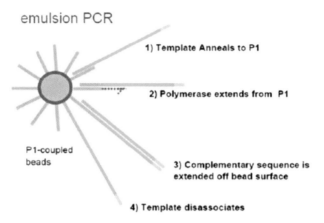

Fig. (9). Amplification of template DNA on bead by emulsion PCR.

4. Discard unamplified beads and use large polystyrene beads coated with P2 sequence to enrich amplified beads, which should contain P2 adaptor.
5. Lay beads on slide
6. Put slides (max of two) in slide chamber
7. Attach the slide chamber to sequencer and make sure the sequencer is ready.
8. Run sequencing

ii) Sequencing run

The design is shown in Fig. (**10**)

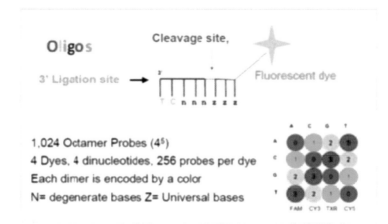

Fig. (10). Design of the SOLiD system.

The chemistry of SOLiD sequencing is shown in (Fig. **11-13**). Basically the procedure is composed of the following steps (Fig. **11**):

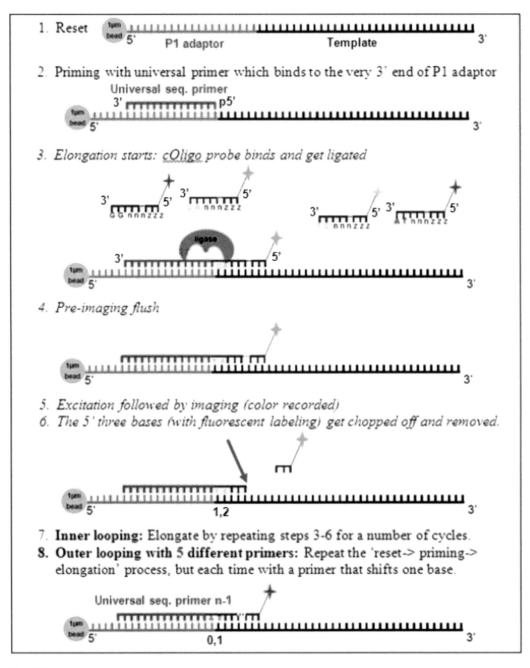

1. Reset

2. Priming with universal primer which binds to the very 3' end of P1 adaptor

3. *Elongation starts: cOligo probe binds and get ligated*

4. *Pre-imaging flush*

5. *Excitation followed by imaging (color recorded)*
6. *The 5' three bases (with fluorescent labeling) get chopped off and removed.*

7. **Inner looping:** Elongate by repeating steps 3-6 for a number of cycles.
8. **Outer looping with 5 different primers:** Repeat the 'reset-> priming-> elongation' process, but each time with a primer that shifts one base.

Fig. (11).

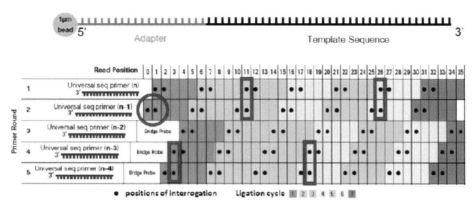

Fig. (12).

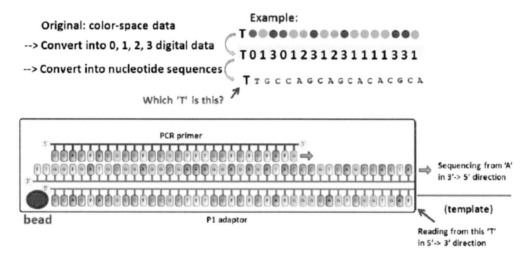

Fig. (13).

1. Reset to ensure that the template is in a clear, single-stranded DNA state
2. Priming by a sequencing primer
3. Elongation: oligo hybridization followed by ligation
4. Pre-imaging flush to clear the background
5. Excitation followed by imaging
6. Chop off the fluorescent signal
7. Inner looping: continue steps 3- 6 till the end
8. Outer looping: continue steps 1-7 to finish the sequencing run.

REFERENCES

Ahmadian, A., Ehn, M., Hober, S. (2006). Pyrosequencing: history, biochemistry and future. *Clin Chim Acta; International Journal of Clinical Chemistry,* Jan;*363*(1-2), 83-94. Epub 2005 Sep 13. [http://dx.doi.org/10.1016/j.cccn.2005.04.038]

Huang, Y.F., Chen, S.C., Chiang, Y.S., Chen, T.H., Chiu, K.P. (2012). Palindromic sequence impedes sequencing-by-ligation mechanism. *BMC Syst. Biol., 6* (Suppl. 2), S10.
[http://dx.doi.org/10.1186/1752-0509-6-S2-S10] [PMID: 23281822]

Mardis, E.R. (2008). Next-generation DNA sequencing methods. *Annu. Rev. Genomics Hum. Genet., 9*, 387-402.
[http://dx.doi.org/10.1146/annurev.genom.9.081307.164359] [PMID: 18576944]

Metzker, M.L. (2010). Sequencing technologies - the next generation. *Nat. Rev. Genet., 11*(1), 31-46.
[http://dx.doi.org/10.1038/nrg2626] [PMID: 19997069]

CHAPTER 4

Genome Assembly, the Genomic Era and the Rise of the Omics Era

Abstract: Genome assembly, or genome sequencing, refers to the process or the end product of sequencing genomic fragments of an organism, followed by piecing together, or 'assembling', in scientific terms, the genomic fragments in sequential order to reveal the original genome sequence. To make a genome assembly usable as a reference, annotation (*i.e.*, assigning locations for genes along the chromosome) is also required. The end product of genome sequencing is a complete set of nucleotide sequence(s), of a genome in linear or circular, of DNA or RNA form, depending on the organismic species. It represents the complete genetic makeup determining all molecular potential, entities and activities of that organism. Similar to road maps used for guiding traffic and for helping people to find a person living at a specific address, genome assemblies act as genomic maps (references) to guide us to find genes (eq. persons), regulatory elements, or mutations in specific locations in the genome. Genome assembly aims to generate a genomic map for future studies of that organism and other related organisms.

Keywords: *De novo* genome assembly, Genome assembly, Genome sequencing, Human Genome Project, Omics era, Resequencing.

Definition of terminologies

de novo genome sequencing/assembly

'de novo' is a Latin expression meaning 'from the beginning' or 'anew'. 'de novo genome assembly', or 'de novo genome sequencing', refers to the sequencing followed by the assembly of a genome which has never been done previously. As one can expect, de novo sequencing is much more difficult than resequencing because the latter already has a copy for comparison.

Resequencing

Genome resequencing refers to the sequencing of a genome which has been previously sequenced. A genome is re-sequenced for various reasons, including for improving sequence reliability, or for identifying individual variations, etc.

INTRODUCTION

All genomes evolve from evolution. Each genome represents a track of evolution. With a complete set of genetic instruction each genome directs the molecular activities associated with the making, the maintenance and the death of an organism, as well as its interaction with other organisms in the environment. A combination of all genomes of all organisms available on Earth represents the outcome of the previous evolutionary process and the genetic blueprint which will guide evolution to shape any future biological field.

PART I. REVIEW OF THE STRATEGIES EMPLOYED FOR GENOME ASSEMBLY

Genome assembly involves technological, molecular and computational strategies to figure out the complete sequence of a genome. Before the automation of Sanger sequencing (see Chapter 1), small DNA sequences were assembled without serious computational algorithms. During that stage, the process relied more on molecular cloning strategies to facilitate the sequencing process. Later, with the advent of sequencing automation (automated Sanger sequencing and next-generation sequencing), sophisticated computational algorithms were developed to deal with large volumes of all kinds of sequence data. A co-evolution between sequencing technologies and computer technologies (hardware and software) thus became a unique feature of the Genomic Era.

A few milestones marked the progress of the Genomic Era. The first genome sequenced was phiX174, a single-stranded circular DNA bacterial phage of 5386 bp in size, by Fred Sanger and colleagues (Sanger 1977, Nature 265:687). In the process, the phage genome was physically mapped by restriction mapping and sequenced by 'plus (viral strand) and minus (complementary strand) method' primed with restriction fragments. Later, a few phages (bacterial viruses) and viruses with genome sizes ranging between a few hundred and a few thousand base pairs were subsequently sequenced and assembled using Sanger sequencing in conjunction with restriction mapping or cosmid cloning.

In 1995, the *H. influenzae* Rd genome of 1,830,137 bp in size was published by Fleischmann and colleagues. It was achieved by whole-genome shotgun sequencing followed by assembly with TIGR assembler implemented with advanced computational methods and algorithms. Besides the shotgun sequences, paired reads were employed to help define the order of contigs and the sizes of gaps within scaffolds (Fleischmann *et al.*, 1995). This work was in fact a pilot project to test the hypothesis that an entire genome of several Mb in size can be sequenced by whole-genome shotgun sequencing and assembled by an integrated assembler. In practice, computational methods were developed to create contigs assemblies from 300 – 500 bp cDNA shotgun sequences and read pairs (eq. paired-ends) were used to create scaffolds, within which the number and the sizes of gaps could be estimated. This project represented a milestone for whole-genome shotgun sequencing.

In 2000, the whole-genome shotgun assembly of the *Drosophila melanogaster genome* of

~120 Mb was published by Myers *et al*. It was achieved by shotgun sequencing of bacterial artificial chromosomes (BACs) and assembling by Celera Assembler (Myers *et al*., 2000).

Technological impacts on genome assembly

A number of factors made critical impacts on the evolution of genome assembly. Here the author would like to emphasize the impacts made by 1) NGS and by 2) oligo-mediated technologies. As you might have already sensed, NGS also heavily relies on oligos (*e.g.*, for *in situ* PCR amplification of templates and for priming sequencing reactions). However, it would make it easier for our discussion if we temporarily treat it as a stand-alone technology founded on its non-oligo-based attributes.

Next-Generation Sequencing (NGS)

In 2005, Jonathan Rothberg invented the 454 sequencer, the first NGS machine, in Connecticut, USA. Simultaneously, he created the terminology "Next-Generation Sequencing", or NGS for short, to distinguish his approach from Sanger sequencing (manual and then automated) used by ABI sequencers manufactured in the San Francisco Bay Area. The subsequent marketing of Solexa and SOLiD sequencers helped to shape up the era of NGS. Notice that NGS employs *in situ* PCR, instead of cosmid or BAC cloning, for clonal amplification of templates and that NGS sequencers abandoned di-deoxynucleotide-based Sanger sequencing method, and claimed a number of advantages including low cost, high speed, high throughput, and high yield. However, short read length is a common problem across all NGS platforms and demands another wave of innovation in assembly algorithms (Miller *et al*., 2010) .

When it comes to genome assembly, a number of factors need to be taken into account. These include, but not limited to, genome size, repetitive sequences, assembler, whether there is/are relative genome(s) already available. The most common problems are the short read length and the repetitive sequences interspersed across the genome and those located in the telomere and centromere regions. Long repetitive sequences are more likely to be present in telomere and centromere regions, while short repetitive sequences are more likely to be present in microsatellites

The procedure of the sequencing library construction for a NGS-based genome assembly turned out to be much simpler than that of using Sanger sequencing. Basically, it contains the following steps: 1) collect genomic DNA (gDNA), 2) sonicate gDNA randomly with a sonicator or non-randomly with restriction enzymes (REs). The former is preferred because no discrimination is involved. 3) select DNA fragments within a specific range by running agarose gel electrophoresis, followed by gel excision with a sharp razor blade and DNA extraction from agarose gel, 4) end-repair, 5) construct sequencing library: ligation to sequencing adaptor either by blunt-end ligation or sticky-end ligation depending on the design, and 6) sequencing.

Currently, the most common approach undertaken is to use NGS sequencers and combine short reads produced by Illumina's sequencers with MP (Mate-Pair) reads produced by the 454 system. Strategies to overcome the repetitive sequence problem include the following: 1) To increase read length from short reads to long reads. Datasets of long reads produced by the 454 system were commonly used to compensate the short read problem. However, the read length of the 454 system is limited by its homopolymer problem. On the other hand, the Illumina system is increasing the read length from less than 100 bp to near 300 bp per read in Paired-End sequencing (total will be close to 600 bp), but still has problems resulting from the intrinsic limitations defined by NGS technologies. In the future, single-molecule sequencing promises to extend the read length up to tens or even hundreds of kilo-bases. However, the error rate remains unacceptably high (~10%) and thus has no marketing value. 2) To improve the algorithms for assemblers. Bioinformaticians have been working on improving the capability of assemblers to enhance the accuracy of contigs and scaffold assembly, although the effect is still limited.

Oligo-mediated technologies

It is worth emphasizing the role played by technologies, especially chemical engineering and biotechnologies, along with the evolution of genome sequencing and assembly because technologies and experimental strategies are intimately interwoven and, in fact, strategies to be taken have to depend upon the technologies available at hand. Here the author would like to give a simple example, which is chemical synthesis of oligonucleotides, or "oligos" in short. Although nowadays an oligo synthesizer is no longer considered as something of importance, its influence cannot be ignored. Oligos are known as single-stranded oligonucleotides of ≤ 50 bases in size which are normally synthesized by being chemically synthesized in 3' to 5' direction. The first direct chemical synthesis of dinucleotide, dithymidinyl, was made possible by Michelson and Todd in 1955, while the fully automation of oligo synthesis was achieved in the 1970s. The availability and customizability of oligos significantly facilitated, in many aspects, the progress of biotechnologies, to name a few, including *in situ* hybridization, PCR (polymerase chain reaction) and, of course, the sequencing technologies.

PART II. THE GENOMIC ERA

The Human Genome Project

The human genome programs our life and death. Two centuries of biological investigations made scientists realize that sequencing and decoding the human genome is critical and inevitable. The Human Genome Project (HGP) aimed to sequence the entire human genome of about 3 Gb in size (see Tables **1** and **2**). The idea was initiated in 1984 by the United States Department of Energy (DOE), and the action was officially launched in 1990. A few draft versions were published along its progress and the finished version was published in 2003.

Table 1. Important events during the pre-genomics era.

Year	Description
1859	**Charles Darwin** published "The Origin of Species" and established the concept of evolution.
1865-	**Mendel** presented his data to the Brunn Society for the Study of Natural Science. Later he published his results in the society's journal and sent a copy to Darwin, who gave no response.
1900-5	Rediscovery of Mendel's discovery: William Bateson translated Mendel's paper into English and coined the term "genetics".
1902	In his paper published in *The Lancet*, Archibald E. Garrod suggested that genes were instructions for making enzymes.
1903	Walter Sutton and Theodor Boveri independently proposed a theory of where the genes can be found in the chromosomes.
1907-	Thomas Morgan started working on Drosophila genetics
1925-	"The Uncertainty Principle" developed by Heisenberg and Bohr
1920s	Two types of nucleic acids were discovered: ribonucleic acid (RNA) and deoxyribonucleic acid (DNA)
1928-1944	The genetic material was found to be DNA, not protein, as demonstrated by Fred Griffith (in 1928) and Oswald Avery (in 1944) through transformation of *Streptococcus pneumoniae* (or Pneumococcus**)**
1953	James Watson and Francis Crick published DNA structure
1958	Cesium chloride gradient ctf. and semiconservative DNA rep. discovered/invented by Meselson and Stahl
1950s-60s	Bacterial phages were intensively studied.
1970s	Tumor viruses were intensively studied.
1975-7	Fred Sanger developed DNA sequencing method
1977	First genome assembly, phiX174 published by Sanger *et al.* (Sanger *et al.*, 1977)
1983	Kary Mullis invented PCR method.

Genomic Era

The Genomic Era can be defined as the period of time in human history when scientists concentrated on developing sequencing technologies, and then applied the sequencing technologies to sequence genomes of various organisms. It started around the early 1970s and is still continuing today, although some set the completion of human genome sequencing in 2003 as its end.

Post Genomic Era

As just mentioned above, some scientists define the completion of human genome sequencing as the end of the Genomic Era. By accepting this definition, the Post Genomic Era starts from 2003 when the human genome sequence, as well as many other genome sequences, became available for scientists to analyze for biological and medical applications.

KEY EVENTS LEADING TO THE DEVELOPMENT OF THE GENOMIC ERA

A number of important events occurred during the past few decades to facilitate the progress of biological studies (Table 1). These events practically fostered the advance of the Genomic Era (Table 2).

Genome projects

Table 2. Development of the Genomics Era.

Year	Description
1984-6	Department of Energy (DOE) and others proposed the idea of sequencing the entire human genome.
1986	Scientists met in Cold Spring Harbor to discuss **the Human Genome Project (HGP)**
1990	**HGP** was officially launched with James Watson as the first director
1992	Craig Venter founded **TIGR** (The Institute for Genomic Research) in Rockville, Maryland.
1994	Francis Collins in charge of the **HGP** (estimated price ~$1-0.5/bp, expected to finish in 2005)
1995	*Haemophilus influenzae* genomic sequence completed/published by TIGR
1995	*Mycoplasma genitaliam* (a parasite in genital track) genome published by TIGR
1996	*Methanococcus jannaschii* (Archaea) genome published by TIGR
1996-	Many other microbial genomes including *Deinococcus radiodurans were sequenced*
1997	TIGR and HGS (Human Genome Sciences) broke up
1997	*Helicobacterpylori* genome published by TIGR
1998	DNA sequencer: ABI's Prizm 377 gel sequencer & 3700 capillary array sequencer; MD's MegaBACE capillary sequencer
1998	Venter announced to sequence human genome in New York Times
1998	Venter & Perkin Elmer (PE) teamed up to found Celera for sequencing the human genome
1999	Sequencing by hybridization (Hacia, 1999) (This is a methodology for discovering new sequence variants)
2000	Genome sequence of *D. melanogaster* was published (Myers *et al.*, 2000)
2001	Initial sequence of the human genome published by International Human Genome Sequencing Consortium (Lander *et al.*, 2001)

(Table 2) contd.....

Year	Description
2001	First draft of human genome sequence published by Celera (Venter *et al.*, 2001)
2002	First draft of mouse genome published by Mouse Genome Sequencing Consortium (Mouse Genome Sequencing *et al.*, 2002)
2003	The so-called complete draft of human genome (hg16) was released.
2004	A Finished version of the human genome sequence published (hg17) (International Human Genome Sequencing, 2004)
2004	First draft of bovine (cow: Bos taurus) genome published
2005-7	454 multiplex pyrosequencing invented (2005) ; Solexa sequencer invented (2006); SOLiD System announced (2007)

GENERAL PROCEDURE FOR RESEQUENCING

Genome resequencing followed steps 1-6 as described for *de novo* genome assembly. Then, the assembly can be performed by using the existing genomic sequences as the reference and conducted with various approaches. For example, one can align each individual sequence against the reference genome if one trusts the existing genome. Alternatively, similar to the *de novo* genome assembly, one can assemble all sequences to form contigs and then align the contigs to the existing reference genome. This all depends on one's preference.

ADVANTAGES OF DIRECT MAPPING AGAINST A REFERENCE (NORMAL) GENOME

Prior to HGP, without chrLoc information, only known sequences (*e.g.* EST) were able to be used to compare with current sequence data.

Before HGP: used virtual DB for mapping, could only study known genes, *e.g.* SAGE

After HGP: started using genome assembly for mapping, able to discover and study novel genes

PART III. THE RISE OF THE OMICS ERA

The Omics Era

In biological terms, "omics" refers to the systematic study of a particular molecular species in the biological system. Such stratification of biological studies takes advantage of the genome-wide, global momentum of various genome projects and has created the Omics Era.

The increasing number of genome assemblies and the advent of novel biotechnologies have widened our scope and made it possible for us to conduct biological investigations in genome-wide, cell-wide, tissue-wide (and further up) fashion. These events produced a combinatorial effect to foster the rise of the Omics Era.

Here are a few omics terms which we will have an increasing chance to come across: **1) Genomics -** The Human Genome Project, together with genome assemblies finished before and after the HGP, provoked a new wave of genomics studies. It is difficult to define a clear boundary for Genomics. In general, Genomics refers to the study of the sequence, structure, and function of genomes and, sometimes, their associated activities. **2) Exomics -** Exome is a collective term for exons and exomics, as a subfield of genomics, is the study of exons. **3) Regulatomics** – Similar to Exomics, Regulatomics is a subset of Genomics which deals with the sequence, structure and function of regulatory elements in the genome. **4) Metagenomics** – This biological field sequences and analyzes combinatorial genomes which may coexist in specific environments such as animal guts, faeces, or even air-conditioning machines. **5) Transcriptomics** – Transcriptomes is a collective term for transcripts. Transcriptomics is the science that focuses on the global study of a complete set of transcripts. **6) Metabolomics** – This field studies metabolites produced during the metabolic process. **7) Proteomics** – Proteins reside in the downstream of the central dogma. Proteomics deals with proteins. **8) Glycoproteomics** – Glycosylation is a common modification for proteins. Glycoproteomics is a subset of Proteomics which deals with the structure and function of glycosylated proteins.

REFERENCES

Fleischmann, R.D., Adams, M.D., White, O., Clayton, R.A., Kirkness, E.F., Kerlavage, A.R., Bult, C.J., Tomb, J.F., Dougherty, B.A., Merrick, J.M. (1995). Whole-genome random sequencing and assembly of Haemophilus influenzae Rd. *Science, 269*(5223), 496-512.
[http://dx.doi.org/10.1126/science.7542800] [PMID: 7542800]

Hacia, J.G. (1999). Resequencing and mutational analysis using oligonucleotide microarrays. *Nat. Genet., 21*(1) (Suppl.), 42-47.
[http://dx.doi.org/10.1038/4469] [PMID: 9915500]

International Human Genome Sequencing Consortium. (2004). Finishing the euchromatic sequence of the human genome. *Nature, 431*(7011), 931-945.
[http://dx.doi.org/10.1038/nature03001] [PMID: 15496913]

Lander, E.S., Linton, L.M., Birren, B., Nusbaum, C., Zody, M.C., Baldwin, J., Devon, K., Dewar, K., Doyle, M., FitzHugh, W., Funke, R., Gage, D., Harris, K., Heaford, A., Howland, J., Kann, L., Lehoczky, J., LeVine, R., McEwan, P., McKernan, K., Meldrim, J., Mesirov, J.P., Miranda, C., Morris, W., Naylor, J., Raymond, C., Rosetti, M., Santos, R., Sheridan, A., Sougnez, C., Stange-Thomann, N., Stojanovic, N., Subramanian, A., Wyman, D., Rogers, J., Sulston, J., Ainscough, R., Beck, S., Bentley, D., Burton, J., Clee, C., Carter, N., Coulson, A., Deadman, R., Deloukas, P., Dunham, A., Dunham, I., Durbin, R., French, L., Grafham, D., Gregory, S., Hubbard, T., Humphray, S., Hunt, A., Jones, M., Lloyd, C., McMurray, A., Matthews, L., Mercer, S., Milne, S., Mullikin, J.C., Mungall, A., Plumb, R., Ross, M., Shownkeen, R., Sims, S., Waterston, R.H., Wilson, R.K., Hillier, L.W., McPherson, J.D., Marra, M.A., Mardis, E.R., Fulton, L.A., Chinwalla, A.T., Pepin, K.H., Gish, W.R., Chissoe, S.L., Wendl, M.C., Delehaunty, K.D., Miner, T.L., Delehaunty, A., Kramer, J.B., Cook, L.L., Fulton, R.S., Johnson, D.L., Minx, P.J., Clifton, S.W., Hawkins, T., Branscomb, E.,

Predki, P., Richardson, P., Wenning, S., Slezak, T., Doggett, N., Cheng, J.F., Olsen, A., Lucas, S., Elkin, C., Uberbacher, E., Frazier, M., Gibbs, R.A., Muzny, D.M., Scherer, S.E., Bouck, J.B., Sodergren, E.J., Worley, K.C., Rives, C.M., Gorrell, J.H., Metzker, M.L., Naylor, S.L., Kucherlapati, R.S., Nelson, D.L., Weinstock, G.M., Sakaki, Y., Fujiyama, A., Hattori, M., Yada, T., Toyoda, A., Itoh, T., Kawagoe, C., Watanabe, H., Totoki, Y., Taylor, T., Weissenbach, J., Heilig, R., Saurin, W., Artiguenave, F., Brottier, P., Bruls, T., Pelletier, E., Robert, C., Wincker, P., Smith, D.R., Doucette-Stamm, L., Rubenfield, M., Weinstock, K., Lee, H.M., Dubois, J., Rosenthal, A., Platzer, M., Nyakatura, G., Taudien, S., Rump, A., Yang, H., Yu, J., Wang, J., Huang, G., Gu, J., Hood, L., Rowen, L., Madan, A., Qin, S., Davis, R.W., Federspiel, N.A., Abola, A.P., Proctor, M.J., Myers, R.M., Schmutz, J., Dickson, M., Grimwood, J., Cox, D.R., Olson, M.V., Kaul, R., Raymond, C., Shimizu, N., Kawasaki, K., Minoshima, S., Evans, G.A., Athanasiou, M., Schultz, R., Roe, B.A., Chen, F., Pan, H., Ramser, J., Lehrach, H., Reinhardt, R., McCombie, W.R., de la Bastide, M., Dedhia, N., Blöcker, H., Hornischer, K., Nordsiek, G., Agarwala, R., Aravind, L., Bailey, J.A., Bateman, A., Batzoglou, S., Birney, E., Bork, P., Brown, D.G., Burge, C.B., Cerutti, L., Chen, H.C., Church, D., Clamp, M., Copley, R.R., Doerks, T., Eddy, S.R., Eichler, E.E., Furey, T.S., Galagan, J., Gilbert, J.G., Harmon, C., Hayashizaki, Y., Haussler, D., Hermjakob, H., Hokamp, K., Jang, W., Johnson, L.S., Jones, T.A., Kasif, S., Kaspryzk, A., Kennedy, S., Kent, W.J., Kitts, P., Koonin, E.V., Korf, I., Kulp, D., Lancet, D., Lowe, T.M., McLysaght, A., Mikkelsen, T., Moran, J.V., Mulder, N., Pollara, V.J., Ponting, C.P., Schuler, G., Schultz, J., Slater, G., Smit, A.F., Stupka, E., Szustakowski, J., Thierry-Mieg, D., Thierry-Mieg, J., Wagner, L., Wallis, J., Wheeler, R., Williams, A., Wolf, Y.I., Wolfe, K.H., Yang, S.P., Yeh, R.F., Collins, F., Guyer, M.S., Peterson, J., Felsenfeld, A., Wetterstrand, K.A., Patrinos, A., Morgan, M.J., de Jong, P., Catanese, J.J., Osoegawa, K., Shizuya, H., Choi, S., Chen, Y.J. International Human Genome Sequencing Consortium. (2001). Initial sequencing and analysis of the human genome. *Nature,* *409*(6822), 860-921. [http://dx.doi.org/10.1038/35057062] [PMID: 11237011]

Miller, J.R., Koren, S., Sutton, G. (2010). Assembly algorithms for next-generation sequencing data. *Genomics,* *95*(6), 315-327. [http://dx.doi.org/10.1016/j.ygeno.2010.03.001] [PMID: 20211242]

Waterston, R.H., Lindblad-Toh, K., Birney, E., Rogers, J., Abril, J.F., Agarwal, P., Agarwala, R., Ainscough, R., Alexandersson, M., An, P., Antonarakis, S.E., Attwood, J., Baertsch, R., Bailey, J., Barlow, K., Beck, S., Berry, E., Birren, B., Bloom, T., Bork, P., Botcherby, M., Bray, N., Brent, M.R., Brown, D.G., Brown, S.D., Bult, C., Burton, J., Butler, J., Campbell, R.D., Carninci, P., Cawley, S., Chiaromonte, F., Chinwalla, A.T., Church, D.M., Clamp, M., Clee, C., Collins, F.S., Cook, L.L., Copley, R.R., Coulson, A., Couronne, O., Cuff, J., Curwen, V., Cutts, T., Daly, M., David, R., Davies, J., Delehaunty, K.D., Deri, J., Dermitzakis, E.T., Dewey, C., Dickens, N.J., Diekhans, M., Dodge, S., Dubchak, I., Dunn, D.M., Eddy, S.R., Elnitski, L., Emes, R.D., Eswara, P., Eyras, E., Felsenfeld, A., Fewell, G.A., Flicek, P., Foley, K., Frankel, W.N., Fulton, L.A., Fulton, R.S., Furey, T.S., Gage, D., Gibbs, R.A., Glusman, G., Gnerre, S., Goldman, N., Goodstadt, L., Grafham, D., Graves, T.A., Green, E.D., Gregory, S., Guigó, R., Guyer, M., Hardison, R.C., Haussler, D., Hayashizaki, Y., Hillier, L.W., Hinrichs, A., Hlavina, W., Holzer, T., Hsu, F., Hua, A., Hubbard, T., Hunt, A., Jackson, I., Jaffe, D.B., Johnson, L.S., Jones, M., Jones, T.A., Joy, A., Kamal, M., Karlsson, E.K., Karolchik, D., Kasprzyk, A., Kawai, J., Keibler, E., Kells, C., Kent, W.J., Kirby, A., Kolbe, D.L., Korf, I., Kucherlapati, R.S., Kulbokas, E.J., Kulp, D., Landers, T., Leger, J.P., Leonard, S., Letunic, I., Levine, R., Li, J., Li, M., Lloyd, C., Lucas, S., Ma, B., Maglott, D.R., Mardis, E.R., Matthews, L., Mauceli, E., Mayer, J.H., McCarthy, M., McCombie, W.R., McLaren, S., McLay, K., McPherson, J.D., Meldrim, J., Meredith, B., Mesirov, J.P., Miller, W., Miner, T.L., Mongin, E., Montgomery, K.T., Morgan, M., Mott, R., Mullikin, J.C., Muzny, D.M., Nash, W.E., Nelson, J.O., Nhan, M.N., Nicol, R., Ning, Z., Nusbaum, C., O'Connor, M.J., Okazaki, Y., Oliver, K., Overton-Larty, E., Pachter, L., Parra, G., Pepin, K.H., Peterson, J., Pevzner, P., Plumb, R., Pohl, C.S., Poliakov, A., Ponce, T.C., Ponting, C.P., Potter, S., Quail, M., Reymond, A., Roe, B.A., Roskin, K.M., Rubin, E.M., Rust, A.G., Santos, R., Sapojnikov, V., Schultz, B., Schultz, J., Schwartz, M.S., Schwartz, S., Scott, C., Seaman, S., Searle, S., Sharpe, T., Sheridan, A., Shownkeen, R., Sims, S., Singer, J.B., Slater, G., Smit, A., Smith, D.R., Spencer, B., Stabenau, A., Stange-Thomann, N., Sugnet, C., Suyama, M., Tesler, G.,

Thompson, J., Torrents, D., Trevaskis, E., Tromp, J., Ucla, C., Ureta-Vidal, A., Vinson, J.P., Von Niederhausern, A.C., Wade, C.M., Wall, M., Weber, R.J., Weiss, R.B., Wendl, M.C., West, A.P., Wetterstrand, K., Wheeler, R., Whelan, S., Wierzbowski, J., Willey, D., Williams, S., Wilson, R.K., Winter, E., Worley, K.C., Wyman, D., Yang, S., Yang, S.P., Zdobnov, E.M., Zody, M.C., Lander, E.S. Mouse Genome Sequencing Consortium. (2002). Initial sequencing and comparative analysis of the mouse genome. *Nature, 420*(6915), 520-562.
[http://dx.doi.org/10.1038/nature01262] [PMID: 12466850]

Myers, E.W., Sutton, G.G., Delcher, A.L., Dew, I.M., Fasulo, D.P., Flanigan, M.J., Kravitz, S.A., Mobarry, C.M., Reinert, K.H., Remington, K.A., Anson, E.L., Bolanos, R.A., Chou, H.H., Jordan, C.M., Halpern, A.L., Lonardi, S., Beasley, E.M., Brandon, R.C., Chen, L., Dunn, P.J., Lai, Z., Liang, Y., Nusskern, D.R., Zhan, M., Zhang, Q., Zheng, X., Rubin, G.M., Adams, M.D., Venter, J.C. (2000). A whole-genome assembly of Drosophila. *Science, 287*(5461), 2196-2204.
[http://dx.doi.org/10.1126/science.287.5461.2196] [PMID: 10731133]

Sanger, F., Air, G.M., Barrell, B.G., Brown, N.L., Coulson, A.R., Fiddes, C.A., Hutchison, C.A., Slocombe, P.M., Smith, M. (1977). Nucleotide sequence of bacteriophage phi X174 DNA. *Nature, 265*(5596), 687-695.
[http://dx.doi.org/10.1038/265687a0] [PMID: 870828]

Venter, J.C., Adams, M.D., Myers, E.W., Li, P.W., Mural, R.J., Sutton, G.G., Smith, H.O., Yandell, M., Evans, C.A., Holt, R.A., Gocayne, J.D., Amanatides, P., Ballew, R.M., Huson, D.H., Wortman, J.R., Zhang, Q., Kodira, C.D., Zheng, X.H., Chen, L., Skupski, M., Subramanian, G., Thomas, P.D., Zhang, J., Gabor Miklos, G.L., Nelson, C., Broder, S., Clark, A.G., Nadeau, J., McKusick, V.A., Zinder, N., Levine, A.J., Roberts, R.J., Simon, M., Slayman, C., Hunkapiller, M., Bolanos, R., Delcher, A., Dew, I., Fasulo, D., Flanigan, M., Florea, L., Halpern, A., Hannenhalli, S., Kravitz, S., Levy, S., Mobarry, C., Reinert, K., Remington, K., Abu-Threideh, J., Beasley, E., Biddick, K., Bonazzi, V., Brandon, R., Cargill, M., Chandramouliswaran, I., Charlab, R., Chaturvedi, K., Deng, Z., Di Francesco, V., Dunn, P., Eilbeck, K., Evangelista, C., Gabrielian, A.E., Gan, W., Ge, W., Gong, F., Gu, Z., Guan, P., Heiman, T.J., Higgins, M.E., Ji, R.R., Ke, Z., Ketchum, K.A., Lai, Z., Lei, Y., Li, Z., Li, J., Liang, Y., Lin, X., Lu, F., Merkulov, G.V., Milshina, N., Moore, H.M., Naik, A.K., Narayan, V.A., Neelam, B., Nusskern, D., Rusch, D.B., Salzberg, S., Shao, W., Shue, B., Sun, J., Wang, Z., Wang, A., Wang, X., Wang, J., Wei, M., Wides, R., Xiao, C., Yan, C., Yao, A., Ye, J., Zhan, M., Zhang, W., Zhang, H., Zhao, Q., Zheng, L., Zhong, F., Zhong, W., Zhu, S., Zhao, S., Gilbert, D., Baumhueter, S., Spier, G., Carter, C., Cravchik, A., Woodage, T., Ali, F., An, H., Awe, A., Baldwin, D., Baden, H., Barnstead, M., Barrow, I., Beeson, K., Busam, D., Carver, A., Center, A., Cheng, M.L., Curry, L., Danaher, S., Davenport, L., Desilets, R., Dietz, S., Dodson, K., Doup, L., Ferriera, S., Garg, N., Gluecksmann, A., Hart, B., Haynes, J., Haynes, C., Heiner, C., Hladun, S., Hostin, D., Houck, J., Howland, T., Ibegwam, C., Johnson, J., Kalush, F., Kline, L., Koduru, S., Love, A., Mann, F., May, D., McCawley, S., McIntosh, T., McMullen, I., Moy, M., Moy, L., Murphy, B., Nelson, K., Pfannkoch, C., Pratts, E., Puri, V., Qureshi, H., Reardon, M., Rodriguez, R., Rogers, Y.H., Romblad, D., Ruhfel, B., Scott, R., Sitter, C., Smallwood, M., Stewart, E., Strong, R., Suh, E., Thomas, R., Tint, N.N., Tse, S., Vech, C., Wang, G., Wetter, J., Williams, S., Williams, M., Windsor, S., Winn-Deen, E., Wolfe, K., Zaveri, J., Zaveri, K., Abril, J.F., Guigó, R., Campbell, M.J., Sjolander, K.V., Karlak, B., Kejariwal, A., Mi, H., Lazareva, B., Hatton, T., Narechania, A., Dicmcr, K., Muruganujan, A., Guo, N., Sato, S., Bafna, V., Istrail, S., Lippert, R., Schwartz, R., Walenz, B., Yooseph, S., Allen, D., Basu, A., Baxendale, J., Blick, L., Caminha, M., Carnes-Stine, J., Caulk, P., Chiang, Y.H., Coyne, M., Dahlke, C., Mays, A., Dombroski, M., Donnelly, M., Ely, D., Esparham, S., Fosler, C., Gire, H., Glanowski, S., Glasser, K., Glodek, A., Gorokhov, M., Graham, K., Gropman, B., Harris, M., Heil, J., Henderson, S., Hoover, J., Jennings, D., Jordan, C., Jordan, J., Kasha, J., Kagan, L., Kraft, C., Levitsky, A., Lewis, M., Liu, X., Lopez, J., Ma, D., Majoros, W., McDaniel, J., Murphy, S., Newman, M., Nguyen, T., Nguyen, N., Nodell, M., Pan, S., Peck, J., Peterson, M., Rowe, W., Sanders, R., Scott, J., Simpson, M., Smith, T., Sprague, A., Stockwell, T., Turner, R., Venter, E., Wang, M., Wen, M., Wu, D., Wu, M., Xia, A., Zandieh, A., Zhu, X. (2001). The sequence of the human genome. *Science,*

291(5507), 1304-1351.
[http://dx.doi.org/10.1126/science.1058040] [PMID: 11181995]

Part 2

Application of DNA Sequencing in Biological Investigations

CHAPTER 5

Laboratory Setup and Fundamental Works

Abstract: This chapter aims to share some previous experiences in laboratory setup and bioinformatics exercises with readers and hopefully, by using this chapter as a mediator, to reduce problems which may be encountered by some readers, especially those who haven't had a chance to personally use sequencers for their studies, and those who have just begun to acquire a taste of sequencing and/or genomics studies. There are many big laboratories and sequencing centers which are able to give you some thoughts and useful opinions. Please consult these resources if possible.

Keywords: Drylab, Sequencing Libraries, Wetlab.

Definition of Terminologies

Sequencing Library

A "sequencing library" is defined as the library actually being sequenced by a sequencer. Since sequencing of an unknown target of DNA molecules has to start from a known and well-defined region, ligation of target DNA molecules to a pair of sequencer-dependent sequencing adaptors is essential for making a sequencing library – also see Chapter 6 for the definition of 'sequencing libraries' and classifications.

Pre-Library

A "pre-library" refers to any type of library built before the sequencing library. It is so defined just to distinguish all the other libraries from the sequencing library. Thus, a pre-library can be the total RNA library, the mRNA library, or the cDNA library used to construct the sequencing library.

Wetlab and Drylab

Here, "wetlab" refers to the lab division involved in the preparation of materials, pre-libraries, sequencing libraries as well as all sorts of bench works in the lab. In contrary to a wetlab, a drylab refers to the lab division involved in data analysis.

INTRODUCTION

Sequencing has myriad applications in many fields, either directly or indirectly related to biology, and its impact on our lives is expected to be enormous and profound. Thanks to those scientists, working either at industry or academia, who have become personally engaged in the progress from manual Sanger sequencing to automated Sanger sequencing, and from automated Sanger sequencing to next-generation sequencing and single-molecule sequencing. Without their efforts, sequencing technologies wouldn't have been able to move forward so expeditiously and sequencing wouldn't be as efficient as we have observed today. However, as shown in the past few decades of human history, sequencing has never been a stand-alone technology and there are many sequencing technologies/platforms and technologies involved, especially computer sciences and biotechnologies.

We have seen a coevolution between sequencing technologies and computer sciences over the years, and this trend is expected to continue for many more decades to come. Without computerization, sequencing automation wouldn't be possible, and without the improvement in speed and capacity of computers and computer-associated devices, the handling of sequence data would have been severely hampered. Furthermore, the development of computer software has added another driving force. Therefore, the role of computer software in genomics investigations is expected to become much more important during the Post Genomic Era.

We have also witnessed a coevolution between sequencing technologies and biotechnologies including those directly or indirectly related to sequencing. In a broad sense, biotechnologies refer to lab methods or methods associated with laboratory equipment. Some lab procedures, such as protein, RNA, or DNA preparations, although also with clear procedures and purposes, are not always recognized as biotechnologies. Some technologies, such as PCR, X-ray crystallography, and sequencing technologies, as each have a well-defined procedure and purpose, making them stand out to be recognized as technologies or biotechnologies, when used for biological investigations. Methods like DNA or RNA isolation may only be considered as laboratory procedures, instead of biotechnologies. However, the fact is, no matter if it is a procedure for DNA or RNA isolation, or a method for molecular cloning for sequencing library construction, it may exert some kind of impact on sequencing efficacy and should not be ignored. For instance, an improved RNA isolation procedure may significantly enhance the quality of a transcriptome library and sequence data. Thus, understanding the rationale and detail of, at least, some key sequencing-related biotechnologies will help you identify the key points which may go wrong, so you can prevent these potential problems from happening by double checking these steps.

Genome sequencing during the genomic era has generated a tremendous amount of sequence data and the data-producing speed is still increasing at an exponential rate during the post genomic era. Undoubtedly, high throughput DNA sequencing will result in a convergence of biology-related fields, asking every biological phenomena and diseases to be traced back to a single nucleotide level, which is then able to provide the highest resolution and has a direct link to genetic mutations and epigenetic modifications.

To cope with the revolutionary situation, a different concept for a laboratory setup and management is desired. First of all, besides a wetlab setup to handle cell culture and experiments (and sequencing in some cases), it is strongly recommended to recruit at least one or more bioinformaticians to work in the lab to handle sequence data, setup and maintain server or computer software and help sequence data analysis.

Here, I list a few things which may be of interest. However, if you are not interested, or are already an experienced lab leader, please skip this chapter.

I. WETLAB

Setting up a Wetlab

The setup of a genomics laboratory is similar to that of a regular research lab, except some may be luckier than others to have sequencers and reagents for their sequencing library construction and sequencing. Some years ago, sequencer manufacturers started to make medium-sized sequencers for medium-sized laboratories, so that we have seen many more labs equipped with these types of sequencers, although not too technical, but workable for certain purposes such as for clinical samples or diagnostics. Very likely this type of sequencer will be as popular as the PCR machine, when everyone in the lab will be able to prepare sequencing libraries and run sequencing independently. For the time being, it is strongly recommended to assign a specialist to take care of the sequencer(s) and sequencing.

Protocols to Use

For a number of reasons, it seems inadequate to describe protocols used for a sequencing library construction in detail, unless under certain special circumstances. First of all, there are a number of protocols currently available for NGS sequencers (Meyer and Kircher, 2010), and each has pros and cons of its own (Head *et al.*, 2014; van Dijk *et al.*, 2014), making it sometimes difficult to choose one from so many. Secondly, protocols used for making a sequencing library are not only library type-dependent, but also sequencer-dependent. As such, before we ask which protocol to use to make a sequencing library, we have to ask which sequencer will be used to sequence the library. Moreover, not only does every sequencer manufacturer tend to make their own protocols for their own sequencers, some reagent providers and sequencer users also join together, making sequencing protocols extremely diversified. We expect to experience a selection process to take place to de-select some less competitive ones. At the same time, competition between sequencers will also cause some protocols to disappear with their attached machines. For all these reasons, I won't try to sell any NGS protocols unless it's for special reasons.

However, it is important to understand the basic structure of some protocols, especially those used for sequencing library construction. Here, I would like to outline and discuss some basic steps to help readers, especially for the beginners, to acquire some basic idea about how to make a sequencing library for a NGS sequencer. For a detailed protocol, please consult your sequencer provider.

In general, the procedure used for constructing a sequencing library can be outlined as below:

A. Prepare cell/tissue samples.
B. Prepare target DNAs to be sequenced.
   This step may involve the making of pre-libraries.
C. Construct sequencing libraries from "pre-libraries".
   A library constructor should pay special attention to the initial amount of material used, and the amounts of intermediates produced during the wetlab process as well as the final quantity, such as the number of templates to be sequenced, *etc.* Quantification is essential for cross-library comparisons. The type of sequencing library has to be decided beforehand. It is also important to make sure the library type can be analyzed with pipelines available in the lab. Otherwise, software applications or programs need to be prepared by the bioinformatics team (drylab).
D. Run sequencing.
   NGS sequencers are state-of-the-art devices, which sequence DNA molecules either by synthesis (*e.g.* Solexa and 454 series) or by ligation (*e.g.* SOLiD series). For enticing the booming sequencing business, sequencer manufacturers may place pre-mature instruments in the market. As such, users are recommended to be cautious before buying an expensive sequencer. A sequencer has to be kept in a cold room with a temperature around 20°C, because it may generate a considerable amount of heat. Rubber caps are normally put underneath the sequencer's table legs to reduce the impact of potential vibration which may occur during a sequencing run.
E. After sequencing run
   Results need to be monitored carefully during sequencing. Used slides should be kept in a buffer solution because you may re-use it for machine calibration or re-run sequencing.

Note: Sequencing reagents should be kept at an appropriate temperature and constantly tracked to monitor the sequencing cost.

II. DRYLAB

Setting Up a Drylab

A. Server setup
   Overall, an in-house or outsourced server system needs to be made available for

data storage, database setup, and sequence data backup. At the same time, methods for data tracking and retrieval and quality checking need to be specified in computer programs. All these are basic requirements and essential for dealing with a large amount of sequence data. Moreover, the wetlab and drylab have to work closely to fully appreciate sequencing and its applications. A medium-sized server system, equipped with over a 100 gigabyte memory and/or a few terabyte HD capacity, has to be made available and usable databases (*e.g.* MySQL, Oracle, and UCSC genome assemblies and annotation databases, *etc.*) need to be downloaded from a public or private domain, into the server system. Since genomics investigations are involved in large amounts of sequence data, it is not possible to use remote reference databases.

Besides, you may also need to identify a high-end server for unexpected usage. For example, a server with a few hundred gigabyte memory and/or a few hundred terabyte HD will do.

B. Design a pipeline for each specific job.

C. Coordinate with wetlab to use a pre-defined naming convention to name libraries.

D. Database setup

MySQL and Oracle are frequently used for data storage. For using MySQL or Oracle, a few things have to be done beforehand. These include, but not limited to, the following: 1) design tables and table2table connections; 2) define schema for each table; 3) assign a server administrator to maintain DB; 4) define privileges for each user; 5) use MySQL or Oracle commands for retrieving data information from the database; 6) all original sequences, no matter good or bad, have to be stored without modification in the original files; 7) secondary sequences and associated information produced during downstream data processing have to be made trackable and be associated with their original sequences.

E. Also keep original sequence data separately to ensure you have the original sequences as backup.

F. Setup a few checkpoints for users to monitor the progress and examine the sequence data quality during data processing.

G. Besides databases, sometimes text files should also be generated as a subsidiary method for data storage.

Sequence Data Processing

"Sequence processing" refers to the 'basic' treatment for all sequence reads. For automation and consistency, sequence processing is normally conducted with purpose-oriented pipelines. Moreover, a main pipeline is frequently divided into a few sub-pipelines based on the library type because the method for data processing can be significantly different between different library types.

In general, a pipeline or sub-pipeline contains the following steps. Results (or data) produced in each step is stored in the database.

1. *Sequence cleanup*
    i. *Selection of high quality sequences using score or QV values.* Low quality sequences are removed and discarded in this step.
    ii. *Removal of contaminant sequences*, including those containing vector sequences, if present, and polyA/T/G/C sequences.
    iii. For Paired-End ditag sequences, *unpaired sequence reads may be removed as well.*
2. *Data re-organization*
    1. Re-organize the sequence data by giving a unique ID to each unique sequence read.
    2. Store the cleanup data in database.
3. *Mapping against the corresponding genome* assembly to associate sequence reads with their genomic/chromosomal locations.
4. *Annotation*
    i. Here, 'annotation' refers to the association of a sequence read with its gene, normally using its genomic/chromosome location as a mediator.
    ii. UCSC annotation database is an example.

Methods Used for Sequence Data Processing and Analysis

"Sequence data analysis" refers to the process used to correlate a batch of sequence reads with their biological attributes to reveal their biological meaning and the cellular status. Analysis can be at the gene/molecular level, or at the pathway or cellular level, and an intensive literature search is required. Frequently, *statistics* approaches are also needed, and you may need to download a number of software applications from the Internet to implement an analytical pipeline.

Normally, one has to decide what method to use for sequence data processing and analysis before a sequencing library is constructed. There are numerous software applications available online for sequence data analysis. SF transcriptome libraries can be analyzed with RNA-Seq approach, while PED sequences can be analyzed with PED pipeline built in-house.

REFERENCES

Head, S.R., Komori, H.K., LaMere, S.A., Whisenant, T., Van Nieuwerburgh, F., Salomon, D.R., Ordoukhanian, P. (2014). Library construction for next-generation sequencing: overviews and challenges. *Biotechniques, 56*(2), 61-64, 66, 68 passim.
[PMID: 24502796]

Meyer, M., Kircher, M. (2010). llumina sequencing library preparation for highly multiplexed target capture and sequencing. *Cold Spring Harb protoc.,* Jun;*2010*(6) pdb.prot5448.
[http://dx.doi.org/10.1101/pdb.prot5448]

van Dijk, E.L., Jaszczyszyn, Y., Thermes, C. (2014). Library preparation methods for next-generation sequencing: tone down the bias. *Exp. Cell Res., 322*(1), 12-20.
[http://dx.doi.org/10.1016/j.yexcr.2014.01.008] [PMID: 24440557]

CHAPTER 6

Sequencing Libraries and Basic Procedure for Sequencing Library Construction

Abstract: A sequencing library is the library prepared to be put on the sequencer for sequencing. As such, sequencing library construction is essential for sequencing. Since sequencing of an unknown target DNA molecules has to start from a known and well-defined region, and every NGS sequencer manufacturer uses unique sequencing primers for its own machines, ligation of target DNA molecules to a pair of adaptors is not only essential but also sequencer-dependent, for the making of a sequencing library. This chapter carries a mission to clarify various types of sequencing libraries and the general procedure for their constructions.

Keywords: Emulsion PCR, ePCR, *in situ* PCR, Mate pair, MP, Paired-end , Paired-end ditag, PE, PED, Sequencing libraries.

INTRODUCTION

There is a saying "garbage in, garbage out". The quality of a sequencing library directly affects the quality of the sequences output from the sequencer, and it is thus required to maintain sequencing libraries at a high quality (Head *et al.*, 2014). Most sequencing labs or sequencing service providers construct sequencing libraries (Meyer and Kircher, 2010), but some don't. To ensure sequence quality, it's strongly recommended for you to construct your own libraries, unless you are not yet familiar with the procedure, or you have a reliable partner to do the job for you.

CLASSIFICATION OF SEQUENCING LIBRARIES

To facilitate communication, it's helpful to classify sequencing libraries into distinguishable categories. Here are three sets of criteria, or bases, summarized from sequencer manufacturers' conventions together with my personal experiences, which I think should be suitable to serve our purpose: A) wetlab technologies and sequencing design, B) cellular origin of the target molecules, and C) gene expression and regulation.

A. Based on wetlab technologies and sequencing design, all non-miRNA molecules of various cellular origins can be constructed into three types of sequencing libraries: 1) *(shotgun) fragment (SF)library*, 2) *Paired-End (PE) library*, and 3) *Paired-End Ditag, or Mate-Paired (PED/ MP), library*. SF sequencing library is the most straightforward sequencing approach which requires only a single sequencing primer.

Because the target DNA molecules can be ligated to sequencing adaptors in either direction, both strands of the target molecule can be read by the sequencing primer. The chromosomal origin and strand can only be identified by mapping (alignment) of the sequence read against the reference genome assembly. PE sequencing library is defined by "forward (F)" and "reverse (R)" sequencing primers used in the sequencer. PE sequencing strategy is one way to increase sequence length if F read and R read have an overlap. The overlap sequence allows us to merge them into a single read. For this application, the inserted (target DNA) length has to be less than the total length (sum) of F and R reads. PED sequencing library has the 5' tag and 3' tag of the same target molecule, and are linked directly through the wetlab procedure to form a ditag prior to the sequencing library construction. Generally speaking, the span, or distance, variation between paired reads of a PED library is much larger than that of a PE library. This is especially true for mRNA-derived libraries. Since miRNAs are small molecules, fragment sequencing is thus the most suitable choice for miRNA sequencing. Molecules over a hundred bp in size, no matter that it is genomic, transcriptomic, ChIP-EM, ChIP-TFBS, or immuno-library, can be sequenced by either fragment, PE, or PED approach.

B. Based on the cellular origin of the target molecules, sequencing libraries can be categorized into 1) *genomic library*, 2) *transcriptome library*, 3) *miRNA library*, 4) *ChIP-mediated epigenetic modification (ChIP-EM) library*, 5) *ChIP-mediated Transcription factor binding site (ChIP-TFBS) library*, 6) *immuno-sequencing library*, and others. Genomic libraries are constructed from genomic DNA fragments (or genomic cDNA fragments if the target organism has a RNA genome). Transcriptome libraries are built from cDNA libraries derived from mRNA transcripts. Similarly, miRNA libraries are constructed from the complementary DNA molecules of miRNAs. Both ChIP-EM and ChIP-TFBS libraries use antibodies (Abs) each of which recognizes a specific protein to enrich chromatin fragments directly or indirectly bound by the Ab-recognized protein. After the removal of proteins, DNA portions can be constructed into a sequencing library and the genomic locations associated with the Ab-recognized protein can be sorted out. Immuno-sequencing libraries can use DNA (or cDNA) from many different sources, depending on research interest and experimental design.

C. Based on gene expression and regulation, sequencing libraries can be split into three major groups: 1) *genomic library* (*e.g.* genomic libraries for genome assembly or the study of recombination or variation of immune genes); 2) *expression library* (*e.g.* mRNA transcriptome and miRNA transcriptome libraries); and 3) *regulation library* (*e.g.* ChIP-EM (epigenetic modification), ChIP-TFBS (transcription factor binding site), and miRNA library). A non-coding miRNA library can be considered as either a transcription library or a regulation library, because miRNAs are transcribed in a way similar to mRNAs and play a role to (negatively) regulate mRNAs.

GENERAL PROCEDURE FOR SEQUENCING LIBRARY CONSTRUCTION

Step I. Libraries Constructed Prior to Sequencing Libraries

Here, upstream wetlab procedure is defined as the bench work conducted prior to sequencing library construction. For example, before a transcriptome sequencing library can be constructed, normally a mRNA and/or cDNA library needs to be constructed; similarly, before a ChIP-EM sequencing library can be constructed, an antibody-enriched ChIP library has to be constructed beforehand, and so on.

DNA prepared during the pre-library stage is fragmented into smaller fragments to fulfill the requirement of NGS sequencers. DNA fragments within a desired size range are selected (normally by gel excision or beads) to make a sequencing library. Such size selection is essential. DNA molecules exceeding the desired length will be likely to tangle with other templates (during cluster generation), causing the sequencer to be unable to distinguish clusters from each other (during sequencing). This problem will result in difficulties in base-calling for tangled clusters, and the so-generated ambiguous reads will be automatically discarded by the sequencer before output. Overall, the yield will be compromised.

1. Fragmentation of DNA, or RNA, by enzymatic reaction or sonication (preferred)
2. End-repair target DNA
3. Size-select target DNA using magnetic beads (Step II and III are interchangeable.)
4. dA-tailing of the size-selected target DNA

Step II. Ligation of Sequencing Adaptors to Target DNA Molecules

Ligation of target DNA to adaptors is essential for all NGS sequencers. Subsequently, the adaptor-ligated target molecules are denatured and subjected to ePCR (for 454 and SOLiD machines) or cluster amplification (for Solexa machines). Notice that, it is important to quantify the amount of DNA molecules whenever possible, and DNA quantification immediately before ligation to sequencing adaptors is critically important, because overcrowded templates will compromise resolution and thus reduce the yield.

1. Ligate (sequencing) adaptors to target DNA molecules using kits recommended by your sequencer manufacturer
2. Purify the ligated DNA
3. Quality evaluation of the ligated constructs using a Bioanalyzer (Fig. **1**).

Note: This step is able to detect the amount of "free" P1 hP2 adaptors in the preparation, but cannot tell whether these adaptors are incorporated into the construct.

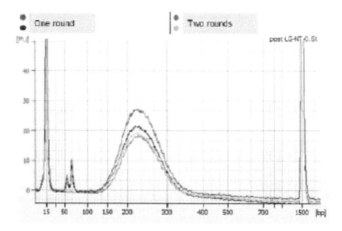

Fig. (1). Bioanalyzer for DNA quality and quantity analysis. Red/deep blue: different samples, both washed only once; deep green/light blue: different samples, both washed twice.

4. Further quantification of the ligated DNA with qPCR to ensure the presence of both adaptors in the construct. This is an important step for us to 'estimate' the amount of P1/hP2-carrying constructs which will really act as templates for emulsion PCR (ePCR).

Step III. Clonal Expansion of Adaptor-ligated DNA Molecules

Here, clonal expansion is defined as the process of building molecular clones for each target DNA molecule so as to lift signals beyond the detectable level during sequencing.

Methods for clonal expansion are sequencer-dependent. Both the 454 and SOLiD systems use off-sequencer emulsion PCR (ePCR) to amplify the "adaptor-target DNA-adaptor" constructs on magnetic beads, while Illumina's instruments eliminate the usage of magnetic beads and generates clusters from individual adaptor-target DNA-adaptor constructs directly onto a sequencing slide. This process is termed cluster amplification. Both ePCR and cluster amplification employ *in situ* PCR to generate clusters, or clones, from well-separated, adaptor-ligated target DNA templates.

A. EMULSION PCR FOR 454 AND SOLID SEQUENCERS

Emulsion PCR is an interesting *in situ* PCR-based technology (Shao *et al.*, 2011). By mixing PCR reagents, pre-prepared in the aqueous phase, with the oil phase, micro-reactors are formed. The process is optimized to maximize the percentage of beads containing only one template and one bead in each micro-reactor, which will lead to a monoclonal amplification. (Micro-reactors with no bead or template will result in empty amplification, micro-reactors with more than one template in each will become poly-clonal, and micro-reactors containing more than two or more beads in each will not only result in a weaker signal but also alter the copy ratio among the target DNA molecules.)

Both the 454 and SOLiD machines adopt emulsion PCR for clonal expansion of adaptor-ligated DNA molecules. Under this strategy, uniform oligos complementary to the adaptors are seated on the magnetic beads. Annealing of adaptor-ligated DNA molecules to the oligos primes the PCR amplification, allowing the synthesis of the other strand to be extended from the oligos. After a number of PCR cycles, each bead is covered by adaptor-ligated DNA molecules. A single denaturation step can remove the unattached template strand.

B. CLUSTER GENERATION

Illumina sequencers (including MiSeq, HiSeq, Solexa, and Genome Analyzer) adopt *in situ* cluster amplification for clonal expansion of the adaptor-ligated DNA molecules (http://www.ucl.ac.uk/wibr/services/solexa/SS_DNAsequencing.pdf). Similar to emulsion PCR, cluster amplification is also an interesting *in situ* PCR-based technology. Cluster amplification uses oligos attached to the sequencing slide as *in situ* PCR primers. Since these oligos are complementary to the adaptor sequences, they anneal to the adaptor-ligated DNA molecules and prime the synthesis of the other strand. After a number of cycles clusters are formed as molecular colonies on the sequencing slide, a simple denaturation followed by enzymatic removing of a particular strand will generate uniform sequences in each cluster and get ready for sequencing.

Apart from HiSeq, Solexa, and GA, the current MiSeq sequencer allows cluster amplification to be conducted directly on the instrument. By doing so, the preparation of a sequencing library is further simplified.

REFERENCES

Head, S.R., Komori, H.K., LaMere, S.A., Whisenant, T., Van Nieuwerburgh, F., Salomon, D.R., Ordoukhanian, P. (2014). Library construction for next-generation sequencing: overviews and challenges. *Biotechniques,* *56*(2), 61-64, 66, 68 passim.
[PMID: 24502796]

Meyer, M., Kircher, M. (2010). llumina sequencing library preparation for highly multiplexed target capture and sequencing. *Cold Spring Harb protoc.,* Jun;*2010*(6) pdb.prot5448.
[http://dx.doi.org/10.1101/pdb.prot5448]

Shao, K., Ding, W., Wang, F., Li, H., Ma, D., Wang, H. (2011). Emulsion PCR: a high efficient way of PCR amplification of random DNA libraries in aptamer selection. *PLoS One,* *6*(9), e24910.
[http://dx.doi.org/10.1371/journal.pone.0024910] [PMID: 21949784]

CHAPTER 7

Paired-End (PE), Mate Pair (MP) and Paired-End Ditag (PED) Sequencing

Abstract: Information regarding the distance between paired reads enhances the accuracy of genome assembly and sequence-to-genome mapping, making paired-end indispensable strategies for DNA sequencing. The most commonly used paired-end sequencing strategies are Paired-End (PE) sequencing and Paired-End Ditag (PED) sequencing. Similarity in terminologies frequently causes confusion. This chapter is set out to clarify these terminologies and then, using PED as an example, to illustrate how a biotechnology can be sequentially developed.

Keywords: bPED, ChIP-EM, ChIP-HM, ChIP-TFBS, *in situ* PCR, mbPED, PE sequencing, PED sequencing.

Definition of Terminologies
PE sequencing: Paired-End sequencing
PED sequencing: Paired-End Ditag sequencing
PET sequencing: Paired-End diTag sequencing

INTRODUCTION

Paired-End diTag (PET) directly links the 5' terminal tags (~18-20 bp each) of genomic DNA fragments or cDNA molecules to their corresponding 3' terminal tags for high throughput sequencing (HTP), has led to a number of important discoveries (Birney *et al.*, 2007; Carninci *et al.*, 2005; Ng *et al.*, 2005; Zhao *et al.*, 2007), including fusion gene identification. To move one-step further, we recently invented a robust method which adopts barcoded adaptors to generate barcoded Paired-End Ditag (bPED) libraries from genomic and transcriptomic libraries. Various bPED libraries, each labeled with a unique internal barcode, can be combined to form a multiplex barcoded Paired-End Ditag (mbPED) library for ultra high-throughput (UHTP) sequencing. These paired-end ditag cloning strategies produce ditag libraries at the lab bench and the ditag libraries can be sequenced as fragment libraries by a single sequencing primer. On the other hand, Paired-End (PE) sequencing is conducted on sequencer using two sequencing primers.

PART I. PE SEQUENCING

Paired-End sequencing uses a forward sequencing primer to sequence the initial set of

Kuo Ping Chiu

template clusters, convert the initial clusters *in situ* into a complementary set of template clusters, and then uses a reverse sequencing primer to sequence the complementary template clusters. Two sets of reads are then paired based on X- and Y-coordinates. Illumina's PE sequencing is a popular sequencing approach for current NGS sequencing.

Illumina PE sequencing includes four steps: *I) library preparation, II) cluster generation, III) sequencing, and IV) data analysis.* Library preparation involves almost all wetlab work, cluster generation and sequencing can now be performed automatically on MiSeq sequencer, and, similar to all other sequencers, data analysis mainly focuses on initial data processing. These steps will be discussed in more detail in the following sections.

I. Library Preparation

The procedure of sequencing library preparation has been a common practice for sequencing labs and sequencing providers. Besides the common procedures for library construction, Illumina also intensify R&D and collaborative effort in developing new protocols to improve accuracy. For related information, please refer to the Sequencing-by-synthesis section of Chapter 2 and Illumina's protocols.

II. Cluster Generation by *in situ* PCR

The amount of DNA required for constructing a sequencing library depends on the type of library to be prepared. For making a transcriptome library, MiSeq requires a total of 600 micro-liter of 12 pM (pico-molar) single-stranded (ss) target DNA. Such diluted concentration was optimized to ensure the separation of templates. About one third is loaded into the chamber of sequencing slide for *in situ* amplification to generate a set of cluster library from which a total of around 15 million raw reads is produced. Thus, the input molecule number to the output cluster ratio is calculated to be about 100 fold excess. Notice that, only the DNA molecules with both ends ligated to sequencing adaptors can be amplified to generate clusters, and the successful rate is another issue.

Prior to cluster amplification, the desired size range has to be determined to prevent cross-over between clusters, which would otherwise impair the specificity of signal. Ambiguous cluster signals are expected to have low quality values and will be discarded eventually. Also, prior to sequencing, the kit of sequencing reagents needs to be determined, so that the sequence length from each primer can be pre-determined. It is better to calculate the overlapped length beforehand.

1. (Two types of oligos, each containing a specific PCR primer, are covalently pre-attached on the surface of a flowcell.) Single-stranded templates, which were already ligated with sequencing adaptors and then denatured, are randomly distributed across the "lawn" of anchored oligos. Optimized concentration allows each template to be well-separated from one another.

2. 1ˢᵗ annealing: Similar to regular PCR reactions, templates are annealed to their complementary template strand

3. 1ˢᵗ elongation: DNA polymerase extends the complementary strand (in "outbound" 5'-to-3' direction) from the PCR primer-defined location until the 5' end of the template

4. 1ˢᵗ denaturation separates the double-stranded structure leaving the oligo-primed strand covalently attached to the surface of the flowcell, while the original templates are washed away from the chamber.

5. 2ⁿᵈ annealing: "Bridge" amplification starts from the second round of PCR because now both PCR primers are all covalently attached on the surface of the flowcell, DNA molecules form "bridges" across the lawn when their free (3') ends anneal to the accessible complementary PCR primers covalently attached to the proximal region of the lawn.

6. 2ⁿᵈ elongation: DNA polymerase makes the complementary strands.

7. 2ⁿᵈ denaturation: denaturation separates the dsDNA molecules and form two ssDNA strands, both of which are covalently attached to the lawn surface at their 5' ends.

8. Continuous PCR cycles eventually generate tens of million copies of clusters on the surface of the flowcell.

9. To achieve a clear sequencing signal, only one strand per cluster is kept for sequencing. This is accomplished by enzymatic digestion of the dsDNA molecules with DNA glycosylaselyase Endonuclease VIII which specifically cleavesU site using USER (Uracil-Specific Excision Reagent; Uracil DNA glycosylase (UDG) + DNA glycosylaselyase Endonuclease VIII). The cleaved strands are washed off the chamber and only the "forward templates" are leave on the lawn for sequencing.

10. Notice that, to prevent unwanted extension from the 3' ends of the sequencing templates. The 3' ends are blocked.

III. Sequencing

Here, an outline is introduced. For detailed procedure please consult Illumina protocol.

Forward Sequencing

1. Before sequencing, clusters are captured tile-by-tile, or panel-by-panel, with camera and their locations "mapped" against X- and Y-coordinates for future referencing to identify the origins of color signals produced during sequencing.

2. Sequencing progresses through "inbound" DNA synthesis primed by sequencing primers bound near the 3' ends of the templates.
Simplification: This inbound sequencing orientation is universally true for sequencing-by-synthesis sequencers, no matter it is conducted inside the flowcell (as adopted by Illumina) or on a magnetic beads (as adopted by Roche 454). This is an intrinsic character defined by the nature of *in situ* PCR amplification for clonal expansion of the

target DNA templates, because the 3' ends of the covalently attached PCR primer oligos have to be made available for outbound PCR elongation, and thus DNA synthesis, which makes the other strand, during sequencing is inbound.

3. Fluorescent signals generated during each cycle are captured by camera tile-by-tile and traced backed to their clusters of origin.

Reverse Sequencing

First, the sequenced strands are stripped off and washed away from the templates used for the first half of the sequencing run.

1. The 3' end blocks are also removed
2. Second run of bridge amplification by *in situ* PCR is performed to make another set of clusters each containing both strands per cluster.
3. The original forward templates are cleaved off by enzymatic reaction by formamidopyrimidine DNA glycosylase (fpg) which cleaves the double-stranded Goxo site on the oligo.
4. The 3' ends of the reverse templates are blocked to prevent unwanted elongation by DNA polymerase during the second half of sequencing
5. Second half of "inbound" sequencing reaction is conducted by using the reverse templates as templates to synthesis the other strands.
6. Locations of captured fluorescent signals are aligned to the focal map and traced back to their clusters of origin.

IV. Data Analysis

1. Similar to all other sequencers, all reads are evaluated based on quality value.
2. Overlapped PE reads are combined to form a "PE-extended" single reads.
3. PE reads without overlap can still be used based on desired situation.

PART II. PAIRED-END DITAG (PED) SEQUENCING

Correction of the Confusion Caused by Inconsistent Terminologies

There was no convention for naming a method, a technology, or a library. Sequencer manufacturers and technology developers tend to name libraries based on their own preferences. This situation can be illustrated by using paired-end ditag as an example. When paired-end ditag technology was first invented in Genome Institute of Singapore (GIS in short), it was named as Gene Identification Signature (GIS), weird enough to confuse people in terminology (Ng *et al.*, 2005). Later, it was, again unconventionally, renamed as Paired-End diTag (PET) (Chiu *et al.*, 2006), and then Paired-End tag (PET), again, silly enough to confuse people. Then, it was called Mate-Paired (MP), a sexy name adopted by Roche. All these miscellaneous terminologies were created just to name the same thing! To avoid further confusion, PED will be used in most occassions to stand for Paired-End Ditag, and PET will be used while we mention the previous approach.

Advantages of PED Sequencing

There are a number of advantages for PED sequencing. **First of all,** as mentioned above, information regarding the distance between paired reads enhances the accuracy of genome assembly and sequence-to-genome alignment. This is an evident advantage of ditag libraries over fragment libraries. **Secondly**, PED strategy is more cost effective and faster (in terms of both sequencing and analysis) than long reads. As such, paired-end sequencing is more suitable for genome-wide investigations. **Thirdly**, PED reads span a broad range from a few hundred to a few thousand bp. **Fourthly**, PED reads well define the boundaries (*i.e.*, the starting location and ending location) of a transcript or a genomic fragment.

Review of PET Technology

PET Cloning Strategy

Technologies come and go like ocean waves. When Paired-End diTag (PET/GIS) was first invented, NGS technologies were not yet available and its cloning strategy followed the conventional/canonical paradigm and was elegantly created by Dr. Ng as shown in (Fig. 1).

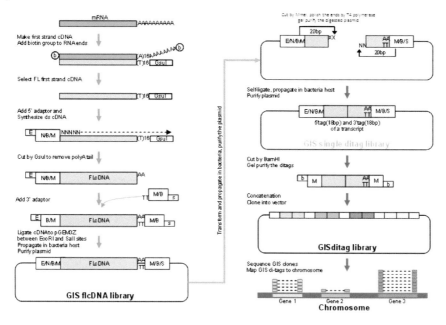

Fig. (1). Cloning Strategy of Paired-End diTag.

The cloning workflow can be summarized into a few steps: 1) mRNA molecules were isolated from total RNA of the cells or tissues under study. 2) Biotin was used to select the 5'CAP+ mRNA molecules, while streptavidin on-beads pull and prevent the 5'CAP+ mRNA molecules from being washed off. 3) (T)16-GsuI oligo was used to select polyA+ mRNA molecules and prime the synthesis of their first strand cDNAs. 4) The first strand cDNA molecules were collected and ligated to the MmeI site-containing 5' adaptor (E-N/B/M-NNNNN), which subsequently primes the synthesis of the second strand cDNAs. 5) the polyA sequence is removed by GsuI digestion, leaving an AA-tail mark in the 3' end. 6) A MmeI site-containing adaptor (M/B-s) is then added to the 3' end. 7) The full-length ds-cDNA molecules were then ligated to, and form a circular structure with, an pre-engineered vector. 8) MmeI digestion releases the internal region to generate paired ditags. 9) Subsequent polishing followed by blunt-end ligation of the wounds (cutting sites) recreated the circular structure. 10) BamHI digestion followed by gel purification generate the ditag structure. 11) Multiple ditags are concatemerized to improve sequencing efficiency. 12) Sequencing produces concatemers.

Identification and Selection of PET Sequences

Right afer sequencing, removing of low quality reads is necessary to avoid or minimize the unnecessary work. Then, ditag pairs were selected by a pipeline (Chiu *et al.*, 2006), which sequentially executed the following steps: 1) Identify the 5' and 3' flanking sequences and the spacers within each concatemer. 2) Collect ditags which reside between the 5' and 3' flanking sequences and spacers. 3) Select quality ditags based on QV and length. 4) Dicard homo-9mer (A_9, T_9, G_9 or C_9)-containing ditags.

PET-to-genome Mapping

Individual tags are mapped against the reference genome assembly, frist directly and then revcom-ed (aiming to generate their complementary strands), against the reference (normal) genome assembly (Fig. **2**). In more detail, 5' tags and 3' tags were first mapped independently so to minimize artificially introduced bias. During the process, individual 5' tags and 3' tags were directly used to map against the plus-strand of each chromosome. (A hit indicates that the tag came from the plus-strand.) Tag sequences were then revcom-ed and mapped against the same chromosome sequences. (A hit indicates that the tag came from the minus-strand.)

The mapped locations of the 5' tag were subsequently paired with the mapped locations of its correspondign 3' tag. A successful pairing has to satisfy the following criteria: The mapped location of the 5' tag and the mapped location of the 3' tag have to be 1) in the same chromosome, 2) on the same strand, 3) in same orientation, 4) in 5' followed by 3' order, and 5) the span (distance between two tags) does not exceed 1 million bp (Fig. **2**).

Based on the pairing result, ditags were split into various categories: 1) MAP0 (no hit), 2)

MAP1 (1 hit per ditag pair), 3) MAP2 (2 hit per ditag pair), 3) MAP3 (3 hit per ditag pair),…, *etc.*

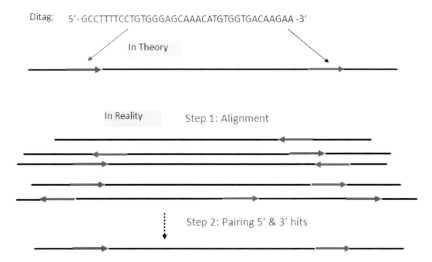

Fig. (2). Mapping and pairing of paired-end ditags.

Application of PET Technology in Transcriptome Analysis

When ditags from a transcriptom library were aligned and clustered along the chromosome, we found them matching very well with the known transcripts collected in the UCSC genome database (Fig. 3). Later, MAP2-MAP4 ditags were found to harbor significant number of pseudo-genes, while the MAP0 ditags contain significant amount of chimeric, or fusion, genes.

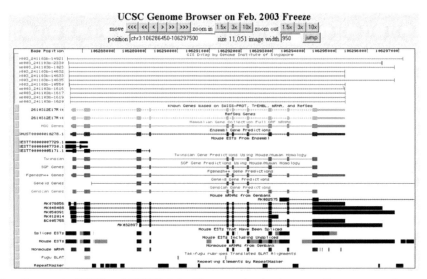

Fig. (3). Co-alignment of PET and UCSC data. The green tracks shown on the upper panel represents ditag spanned region. Notice that the 5' tag and 3' tag are indicated by short bars at the ends. Right below the ditag tracks are gene tracks retrieved from UCSC database.

Application of PET Technology in ChIP-TFBS and ChIP-HM Analyses

After transcriptome analysis, PET technology was subsequently applied to transcription factor binding site (TFBS) and epigenetic modifications (EM) analyses, which were, and still are, mediated by chromatin-immunoprecipitation (ChIP) and was coined as ChIP-PET analysis.

To best fit the purpose, the experimental procesure was modified and finalized as shown in (Fig. 4).

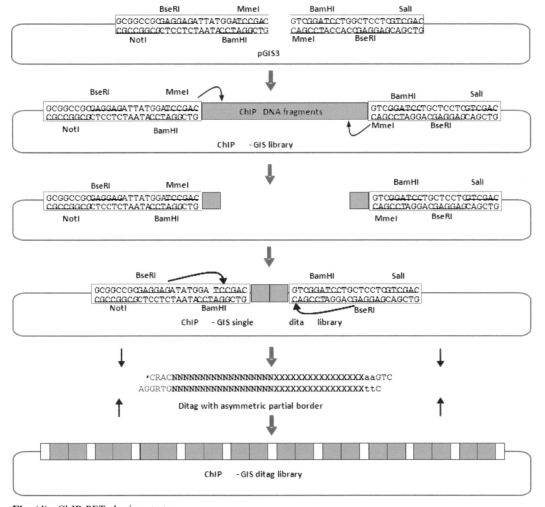

Fig. (4). ChIP-PET cloning strategy.

The results can be examplified by the study published by Zhao and colleagues who reported limited, discrete H3K4 trimethylation, together with broader-scale, continuous H3K27 trimethylation within homeobox domains.

Barcoded Paired-End Ditag and Multiplex Barcoded Paired-End Ditag

PET is a great technology, but no barcode has ever been directly used in the PET construct, and its cloning procedure is tedious. To move one step further, we invented barcoded Paired-End Ditag (bPED) and multiplex barcoded Paired-End Ditag (mbPED) approaches for integrative genomic/transcriptomic/epigenomic investigations.

Here, it would be helpful to clarify a couple of points.

First, bPED does not equal to PET.

The bPED technology uses barcodes (while PET did not) and its cloning strategy is totally different from that of PET (see below).

Second, bPED does not equal to bPE (barcoded PE sequencing).

As we have already disccussed in the beginning of this chapter, there are a number of differences between PED and PE sequencing. These include: 1) bPED is done on wetlab bench before the sequencing library is applied to the sequencer, 2) distance between paired reads can go beyond a few Kb, 3) bPED has both structural and biological meanings, and 4) bPED is constructed and sequenced as fragment library. In contrast, bPE sequencing is determined in sequencer by using F (R1) and R (R2) sequencing primers, and the distance between paired reads of PE sequencing is limited within a few hundred bp, and 3) PE has no structural or biological meaning.

Cloning Strategies of bPED and mbPED

As mentioned above, we added a barcode to each sequencing library. With the labeling of barcodes, multiple bPED libraries can be combined into a mbPED master library. Different from all other existing cloning methods, the barcode of bPED is engineered to lie in the center of the construct (Figs. 5-7).

For more detailed information, please refer to 1) **US patent 8481699. Kuo Ping Chiu (2013).** Multiplex barcoded Paired-End ditag (mbPED) library construction for ultra high throughput sequencing; and 2) **US patent** 8829172**. Kuo Ping Chiu *et al.* (2014). Novel multiplex barcoded Paired-End Ditag (mbPED) sequencing approach and its application in fusion gene identification.**

Thus, multiple (types) of bPED can be combined to generate a mbPED library. For example, for the study of gene expression and regulation, A mbPED library may contain 1) transcriptome library, 2) ChIP-TFBS library, 3) ChIP-EPI library, 4) miRNA library, or 5) multiple single cell transcriptome libraries, *etc.* for a systemic, integrative biological analysis.

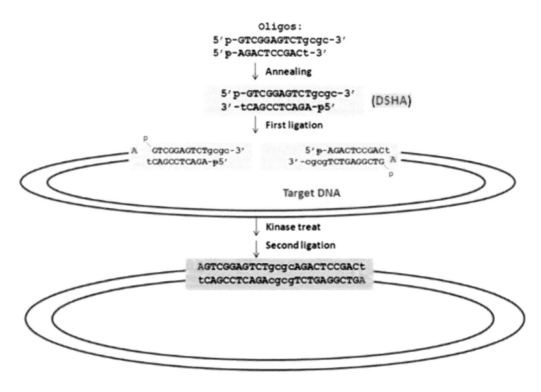

Fig. (5). Preparation of cirlarized construct for making bPED library.

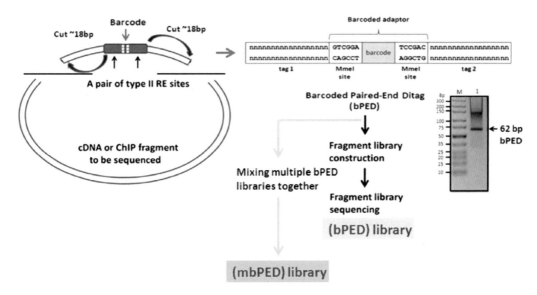

Fig. (6). Generation of bPED and mbPED libraries.

Fig. (7). Multiplex barcoded Paired-End Ditag (mbPED).

Applications of bPED and mbPED

In general, the applications of bPED/mbPED are similar to that of PET, although mbPED is designed for multi-layered, multi-directional ultra high throughput sequencing for integrative biological investigations. bPED and mbPED are suitable for efficient identification of fusion genes from multiple types of cancer cells (Fig. **8**). There are a number of methods already available for fusion gene identification. bPED/mbPED has an evident advantage over the existing approaches. That is, bPED/mbPED approach is ultra high throughput which can screen a number of libraries simultaneously.

Advantages of bPED and mbPED Comparing to PET

There are a number of advantages of bPED/mbPED over the previous PET technology. These include: 1) Procedure is very simple and straightforward. 2) Both first and second ligations are directional. 3) Library construction is completed before sequencing. 4) "All" bPED/mbPED libraries are made as fragment libraries, which can be sequenced by a single sequencing primer. 5) Homogeneous DSHA cannot self-ligate and high ligation efficiency is guaranteed. 6) Only one full adaptor is formed in the circularized construct. 7) Multiple types of libraries can be combined for multi-layered, multi-dimensional systematic sequence analysis. 8) Save sequencing cost and time.

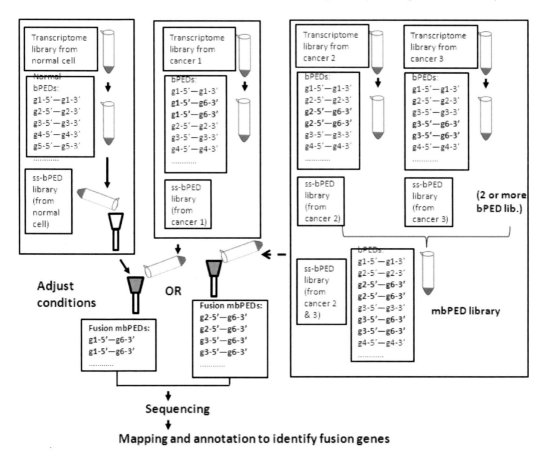

Fig. (8). Application of mbPED in identification of fusion genes.

Fig. (9). Double-stranded secondary structure formed in palindromic region fails hybridization and thus impedes sequencing-by-ligation mechanism.

Formation of an Interior Palindrome is a Unique Feature of bPED

With bPED technology, a palindrome is created in the centralized adaptor region. The presence of palindrome does not have any effect on the efficacy of sequencing-b-

-synthesis-based sequencers. Contrarily, we found that sequencing-by-ligation sequencers (*e.g.*, SOLiD machines) cannot read through the palindromic region (Fig. **9**) (Huang *et al.*, 2012). This is understandable because sequencing-by-synthesis-based sequencers adopt RNA polymerase which has helicase activity for synthesis. On the other hand, sequencing-by-ligation-based sequencers use hybridization-followed-by-ligation approach. Palindromic sequences tend to form (two) double-stranded secondary structures which are able to impair hybridization. Failure in hybridization fails the sequencing.

REFERENCES

Birney, E., Stamatoyannopoulos, J.A., Dutta, A., Guigó, R., Gingeras, T.R., Margulies, E.H., Weng, Z., Snyder, M., Dermitzakis, E.T., Thurman, R.E., Kuehn, M.S., Taylor, C.M., Neph, S., Koch, C.M., Asthana, S., Malhotra, A., Adzhubei, I., Greenbaum, J.A., Andrews, R.M., Flicek, P., Boyle, P.J., Cao, H., Carter, N.P., Clelland, G.K., Davis, S., Day, N., Dhami, P., Dillon, S.C., Dorschner, M.O., Fiegler, H., Giresi, P.G., Goldy, J., Hawrylycz, M., Haydock, A., Humbert, R., James, K.D., Johnson, B.E., Johnson, E.M., Frum, T.T., Rosenzweig, E.R., Karnani, N., Lee, K., Lefebvre, G.C., Navas, P.A., Neri, F., Parker, S.C., Sabo, P.J., Sandstrom, R., Shafer, A., Vetrie, D., Weaver, M., Wilcox, S., Yu, M., Collins, F.S., Dekker, J., Lieb, J.D., Tullius, T.D., Crawford, G.E., Sunyaev, S., Noble, W.S., Dunham, I., Denoeud, F., Reymond, A., Kapranov, P., Rozowsky, J., Zheng, D., Castelo, R., Frankish, A., Harrow, J., Ghosh, S., Sandelin, A., Hofacker, I.L., Baertsch, R., Keefe, D., Dike, S., Cheng, J., Hirsch, H.A., Sekinger, E.A., Lagarde, J., Abril, J.F., Shahab, A., Flamm, C., Fried, C., Hackermüller, J., Hertel, J., Lindemeyer, M., Missal, K., Tanzer, A., Washietl, S., Korbel, J., Emanuelsson, O., Pedersen, J.S., Holroyd, N., Taylor, R., Swarbreck, D., Matthews, N., Dickson, M.C., Thomas, D.J., Weirauch, M.T., Gilbert, J., Drenkow, J., Bell, I., Zhao, X., Srinivasan, K.G., Sung, W.K., Ooi, H.S., Chiu, K.P., Foissac, S., Alioto, T., Brent, M., Pachter, L., Tress, M.L., Valencia, A., Choo, S.W., Choo, C.Y., Ucla, C., Manzano, C., Wyss, C., Cheung, E., Clark, T.G., Brown, J.B., Ganesh, M., Patel, S., Tammana, H., Chrast, J., Henrichsen, C.N., Kai, C., Kawai, J., Nagalakshmi, U., Wu, J., Lian, Z., Lian, J., Newburger, P., Zhang, X., Bickel, P., Mattick, J.S., Carninci, P., Hayashizaki, Y., Weissman, S., Hubbard, T., Myers, R.M., Rogers, J., Stadler, P.F., Lowe, T.M., Wei, C.L., Ruan, Y., Struhl, K., Gerstein, M., Antonarakis, S.E., Fu, Y., Green, E.D., Karaöz, U., Siepel, A., Taylor, J., Liefer, L.A., Wetterstrand, K.A., Good, P.J., Feingold, E.A., Guyer, M.S., Cooper, G.M., Asimenos, G., Dewey, C.N., Hou, M., Nikolaev, S., Montoya-Burgos, J.I., Löytynoja, A., Whelan, S., Pardi, F., Massingham, T., Huang, H., Zhang, N.R., Holmes, I., Mullikin, J.C., Ureta-Vidal, A., Paten, B., Seringhaus, M., Church, D., Rosenbloom, K., Kent, W.J., Stone, E.A., Batzoglou, S., Goldman, N., Hardison, R.C., Haussler, D., Miller, W., Sidow, A., Trinklein, N.D., Zhang, Z.D., Barrera, L., Stuart, R., King, D.C., Ameur, A., Enroth, S., Bieda, M.C., Kim, J., Bhinge, A.A., Jiang, N., Liu, J., Yao, F., Vega, V.B., Lee, C.W., Ng, P., Shahab, A., Yang, A., Moqtaderi, Z., Zhu, Z., Xu, X., Squazzo, S., Oberley, M.J., Inman, D., Singer, M.A., Richmond, T.A., Munn, K.J., Rada-Iglesias, A., Wallerman, O., Komorowski, J., Fowler, J.C., Couttet, P., Bruce, A.W., Dovey, O.M., Ellis, P.D., Langford, C.F., Nix, D.A., Euskirchen, G., Hartman, S., Urban, A.E., Kraus, P., Van Calcar, S., Heintzman, N., Kim, T.H., Wang, K., Qu, C., Hon, G., Luna, R., Glass, C.K., Rosenfeld, M.G., Aldred, S.F., Cooper, S.J., Halees, A., Lin, J.M., Shulha, H.P., Zhang, X., Xu, M., Haidar, J.N., Yu, Y., Ruan, Y., Iyer, V.R., Green, R.D., Wadelius, C., Farnham, P.J., Ren, B., Harte, R.A., Hinrichs, A.S., Trumbower, H., Clawson, H., Hillman-Jackson, J., Zweig, A.S., Smith, K., Thakkapallayil, A., Barber, G., Kuhn, R.M., Karolchik, D., Armengol, L., Bird, C.P., de Bakker, P.I., Kern, A.D., Lopez-Bigas, N., Martin, J.D., Stranger, B.E., Woodroffe, A., Davydov, E., Dimas, A., Eyras, E., Hallgrímsdóttir, I.B., Huppert, J., Zody, M.C., Abecasis, G.R., Estivill, X., Bouffard, G.G., Guan, X., Hansen, N.F., Idol, J.R., Maduro, V.V., Maskeri, B., McDowell, J.C., Park, M., Thomas, P.J., Young, A.C., Blakesley, R.W., Muzny, D.M., Sodergren, E., Wheeler, D.A., Worley, K.C., Jiang, H., Weinstock, G.M., Gibbs, R.A., Graves, T., Fulton, R., Mardis, E.R.,

Wilson, R.K., Clamp, M., Cuff, J., Gnerre, S., Jaffe, D.B., Chang, J.L., Lindblad-Toh, K., Lander, E.S., Koriabine, M., Nefedov, M., Osoegawa, K., Yoshinaga, Y., Zhu, B., de Jong, P.J. ENCODE Project Consortium; (2007). NISC Comparative Sequencing Program; (2007). Baylor College of Medicine Human Genome Sequencing Center; (2007). Washington University Genome Sequencing Center; (2007). Broad Institute; (2007). Children's Hospital Oakland Research Institute. (2007). Identification and analysis of functional elements in 1% of the human genome by the ENCODE pilot project. *Nature, 447*(7146), 799-816. [http://dx.doi.org/10.1038/nature05874] [PMID: 17571346]

Carninci, P., Kasukawa, T., Katayama, S., Gough, J., Frith, M.C., Maeda, N., Oyama, R., Ravasi, T., Lenhard, B., Wells, C., Kodzius, R., Shimokawa, K., Bajic, V.B., Brenner, S.E., Batalov, S., Forrest, A.R., Zavolan, M., Davis, M.J., Wilming, L.G., Aidinis, V., Allen, J.E., Ambesi-Impiombato, A., Apweiler, R., Aturaliya, R.N., Bailey, T.L., Bansal, M., Baxter, L., Beisel, K.W., Bersano, T., Bono, H., Chalk, A.M., Chiu, K.P., Choudhary, V., Christoffels, A., Clutterbuck, D.R., Crowe, M.L., Dalla, E., Dalrymple, B.P., de Bono, B., Della Gatta, G., di Bernardo, D., Down, T., Engstrom, P., Fagiolini, M., Faulkner, G., Fletcher, C.F., Fukushima, T., Furuno, M., Futaki, S., Gariboldi, M., Georgii-Hemming, P., Gingeras, T.R., Gojobori, T., Green, R.E., Gustincich, S., Harbers, M., Hayashi, Y., Hensch, T.K., Hirokawa, N., Hill, D., Huminiecki, L., Iacono, M., Ikeo, K., Iwama, A., Ishikawa, T., Jakt, M., Kanapin, A., Katoh, M., Kawasawa, Y., Kelso, J., Kitamura, H., Kitano, H., Kollias, G., Krishnan, S.P., Kruger, A., Kummerfeld, S.K., Kurochkin, I.V., Lareau, L.F., Lazarevic, D., Lipovich, L., Liu, J., Liuni, S., McWilliam, S., Madan Babu, M., Madera, M., Marchionni, L., Matsuda, H., Matsuzawa, S., Miki, H., Mignone, F., Miyake, S., Morris, K., Mottagui-Tabar, S., Mulder, N., Nakano, N., Nakauchi, H., Ng, P., Nilsson, R., Nishiguchi, S., Nishikawa, S., Nori, F., Ohara, O., Okazaki, Y., Orlando, V., Pang, K.C., Pavan, W.J., Pavesi, G., Pesole, G., Petrovsky, N., Piazza, S., Reed, J., Reid, J.F., Ring, B.Z., Ringwald, M., Rost, B., Ruan, Y., Salzberg, S.L., Sandelin, A., Schneider, C., Schönbach, C., Sekiguchi, K., Semple, C.A., Seno, S., Sessa, L., Sheng, Y., Shibata, Y., Shimada, H., Shimada, K., Silva, D., Sinclair, B., Sperling, S., Stupka, E., Sugiura, K., Sultana, R., Takenaka, Y., Taki, K., Tammoja, K., Tan, S.L., Tang, S., Taylor, M.S., Tegner, J., Teichmann, S.A., Ueda, H.R., van Nimwegen, E., Verardo, R., Wei, C.L., Yagi, K., Yamanishi, H., Zabarovsky, E., Zhu, S., Zimmer, A., Hide, W., Bult, C., Grimmond, S.M., Teasdale, R.D., Liu, E.T., Brusic, V., Quackenbush, J., Wahlestedt, C., Mattick, J.S., Hume, D.A., Kai, C., Sasaki, D., Tomaru, Y., Fukuda, S., Kanamori-Katayama, M., Suzuki, M., Aoki, J., Arakawa, T., Iida, J., Imamura, K., Itoh, M., Kato, T., Kawaji, H., Kawagashira, N., Kawashima, T., Kojima, M., Kondo, S., Konno, H., Nakano, K., Ninomiya, N., Nishio, T., Okada, M., Plessy, C., Shibata, K., Shiraki, T., Suzuki, S., Tagami, M., Waki, K., Watahiki, A., Okamura-Oho, Y., Suzuki, H., Kawai, J., Hayashizaki, Y. FANTOM Consortium; (2005). RIKEN Genome Exploration Research Group and Genome Science Group (Genome Network Project Core Group). (2005). The transcriptional landscape of the mammalian genome. *Science, 309*(5740), 1559-1563. [http://dx.doi.org/10.1126/science.1112014] [PMID: 16141072]

Chiu, K.P., Wong, C.H., Chen, Q., Ariyaratne, P., Ooi, H.S., Wei, C.L., Sung, W.K., Ruan, Y. (2006). PET-Tool: a software suite for comprehensive processing and managing of Paired-End diTag (PET) sequence data. *BMC Bioinformatics, 7*, 390. [http://dx.doi.org/10.1186/1471-2105-7-390] [PMID: 16934139]

Huang, Y.F., Chen, S.C., Chiang, Y.S., Chen, T.H., Chiu, K.P. (2012). Palindromic sequence impedes sequencing-by-ligation mechanism. *BMC Syst. Biol., 6* (Suppl. 2), S10. Epub 2012 Dec 12. [http://dx.doi.org/10.1186/1752-0509-6-S2-S10] [PMID: 23281822]

Ng, P., Wei, C.L., Sung, W.K., Chiu, K.P., Lipovich, L., Ang, C.C., Gupta, S., Shahab, A., Ridwan, A., Wong, C.H., Liu, E.T., Ruan, Y. (2005). Gene identification signature (GIS) analysis for transcriptome characterization and genome annotation. *Nat. Methods, 2*(2), 105-111. [http://dx.doi.org/10.1038/nmeth733] [PMID: 15782207]

Zhao, X.D., Han, X., Chew, J.L., Liu, J., Chiu, K.P., Choo, A., Orlov, Y.L., Sung, W.K., Shahab, A., Kuznetsov, V.A., Bourque, G., Oh, S., Ruan, Y., Ng, H.H., Wei, C.L. (2007). Whole-genome mapping of histone H3 Lys4 and 27 trimethylations reveals distinct genomic compartments in human embryonic stem cells. *Cell Stem Cell, 1*(3), 286-298.
[http://dx.doi.org/10.1016/j.stem.2007.08.004] [PMID: 18371363]

.

CHAPTER 8

Genome Sequencing and Assembly

Abstract: In chapter 4, we reviewed the background and history of genome sequencing and assembly. In this chapter, we will focus more on the technical and experimental issues. In science, the term "whole genome sequencing and assembly" is often used interchangeably with "genome sequencing and assembly", "genome sequencing" and "genome assembly", because genome assembly normally refers to the assembly of an entire genome and genome sequencing is normally followed by assembling sequence reads to produce a complete set of chromosomal sequences for the genome of interest. For convenience, genome assembly and genome sequencing are preferred and will be used more often than the others throughout our discussion. Since whole genome sequencing and assembly is a complicate process, involving multiple alternatives and methodologies, it is not possible to cover every detail. We will go through some concepts and NGS-associated procedures so that readers can get some idea of how the genome assembly is achieved. Serious readers are recommended to consult previous reports published by sequencing laboratories.

Keywords: Contig, *De novo* genome assembly, Genome assembly, Genome sequencing, Scaffold.

INTRODUCTION

The beginning of the 21st century was marked by numerous genome releases. In 1992, Craig Venter founded the Institute for Genomic Research (TIGR) to sequence microbial genomes using whole-genome shotgun sequencing strategy. In 1998, he founded another organization, Celera Genomics, to sequence the human genome using techniques his team developed. Since then, a number of genomes have been sequenced and assembled. These assembled genomes are available in many famous genome centers across the world.

Among the most commonly used reference genomes are the human genome and the mouse genome, both of which were published during the first few years of the 21st century. Key publications include 1) Initial sequencing and analysis of the human genome (Lander *et al.*, 2001); 2) The sequence of the human genome (Venter *et al.*, 2001); 3) Initial sequencing and comparative analysis of the mouse genome (Mouse Genome Sequencing *et al.*, 2002); 4) Finishing the euchromatic sequence of the human genome (International Human Genome Sequencing, 2004); and 5) Quality assessment of the human genome sequence (Schmutz *et al.*, 2004).

Besides, both *Drosophila melanogaster* and *Arabidopsis thaliana* genomes were published in 2000 (Arabidopsis Genome, 2000; Myers *et al.*, 2000), while the *C. elegans* genome was first released in 1998 (as the first animal genome release) (Scienc 282:2012-2018) and re-sequenced in 2008 (Hillier *et al.*, Nature Methods 5:183).

The haploid human genome consists of an approximately 3 Gb sequence distributed in 23 nuclear chromosomes and a mitochondrial chromosome. About 20,000 - 25,000 protein-coding genes and at least hundreds of noncoding genes involved in post-transcriptional regulation have been found in the human genome. This kind of sequence information can be best acquired by whole genome sequencing coupled with whole transcriptome sequencing. Genome assembly aims to build a genetic map to be used as a reference genome for future biological studies. Moreover, reference genome sequences can be used to facilitate the assembly of related genomes.

CLASSIFICATION OF GENOME ASSEMBLY

In general, genome assembly can be categorized into *de novo* genome assembly (or *de novo* genome sequencing) and re-sequencing. **De novo genome assembly** refers to the sequencing of the DNA or RNA genome of an organism which has never been sequenced before, followed by using assembler software or by implementing a processing pipeline to piece together the sequence reads, so to generate a "genomic map" for that target organism. Then, sequence-to-gene annotation is conducted to associate chromosomal locations with genes and other genomic entities such as enhancers, promoters, DNA motifs, *etc*. Such assembled genome can then be used in the future for the studies of that organism as well as its related species. On the other hand, **re-sequencing** refers to the sequencing of a genome of an individual, of which a reference genome has previously been built. With the increasing number of genomes having been assembled, more and more "relative genomes" are becoming available, and these relative genomes can be used as references to facilitate the assembly process of a similar genome. This type of genome assembly can be categorized as **"semi-*de novo* genome assembly"**.

WHAT SEQUENCING STRATEGIES TO TAKE?

Among the NGS machines, a combination of 454 (*e.g.* Flex) with Illumina's sequencers (*e.g.* HiSeq series) is frequently used for genome assembly. This strategy takes advantage of 454's long reads and HiSeq's high quality and high yield. Usually, paired-end ditag, paired-end, and, sometimes, fragment libraries are combined to achieve the accuracy of the assembly process. (SOLiD machines, which use sequencing-by-ligation mechanism, are not suitable for genome assembly, simply because these machines are incapable of reading through palindromic structures.)

GENERAL PROCEDURE FOR *DE NOVO* GENOME SEQUENCING

Strategies for genome assembly have experienced a dramatic change since the launch of

Human Genome Project in 1990. The major technical impact resulted from the introduction of NGS technologies starting from 2005. The revolutionary NGS technologies were later accompanied by a wave of software evolution attempting to deal with the ever-increasing huge amounts of short read sequence data produced by various NGS machines. To make a long story short, here I would like to use the shotgun sequencing strategy, which is commonly taken by scientists working on genome assembly, as an example to illustrate how a genome can be assembled (Fig. 1).

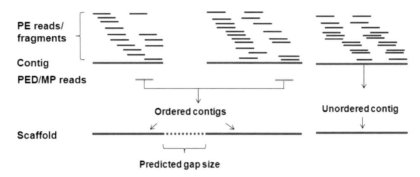

Fig. (1). The first part of *de novo* genome assembly. Sequence reads, either paired-end or fragment reads, are assembled into contigs, which are then ordered and assembled into scaffolds. Based on additional information, gaps between ordered contigs can be predicted.

The general procedure for *de novo* genome assembly is outlined below:

A. ***Preparation of sequencing libraries***
  1. Isolate genomic DNA (gDNA) from an organism. White blood cells in the blood is frequently used for gDNA isolation. There are a number of kits/methods that can be used to isolate genomic DNA from prokaryotic or eukaryotic genome. To obtain these commercial products or protocols, please consult experienced personnel, Maniatis Manual, or use Google search. Once you have genomic DNA ready, you can continue to make a few types of sequencing libraries for genome assembly. The common ones include fragment library for shotgun short-read sequencing and Paired-End library for ditag sequencing.
  DNA quantification
  Keep in mind to trace the quantity of your sample frequently. Do quantify your DNA before and after sonication, and before *in situ* PCR amplification. Initial DNA quantification, which does not have to be very accurate, can use NanoDrop or bioanalyzer.Normally, there is no difficulty to obtain a sufficient amount of genomic DNA, because a quantity at only a micro-gram scale is required for each library. More accurate quantification of DNA can be done by bioanalyzer or fragment analyzer which display DNA molecules based on size. The most accurate quantification can be done by qPCR, if necessary.
  2. Sonicate gDNA to make gDNA fragments.

DNA samples are sonicated into small fragments of a few hundred bp in size. It is recommended to use the Covaris system or probe for DNA sonication. The sonication conditions have been tested for shearing 10 ng – 5 g DNA in a one hundred to a few hundred L. For certain NDA samples, optimizing the shearing protocol may be necessary.

3. Select gDNA fragments with a few hundred bp in size.

Fragments of a few hundred base pairs are desired and isolated by gel excision and then ligated to sequencing adaptors for sequencing library preparation.

4. End-repair (End-polish)

Sonication may generate 5′ overhang, 3′ overhang, or blunt end for a DNA fragment. Procedure of end-repair uses DNA polymerase to 1) fillin the complementary strand for 5' overhand, 2) trim off 3' overhang, and 3) leave blunt-end intact.

5. A-tailing

This step adds an 'A' base to the 3'end of all DNA fragments. Depending on the design of sequencing adaptors, some protocols ask you to tail the size-selected and end-polished DNA molecules with an extra A base to the 3' ends. A thermostable polymerase adds an A to each 3′ end of the non-templated DNA strand. The thermostable polymerase lacks 3′ to 5′ exonuclease activity at higher temperatures.

6. Ligate the selected gDNA fragments to sequencing adaptors.

Quantificatify DNA bioanalyzer or fragment analyzer after ligation. (Bioanalyzer is much more expensive than fragment analyzer.)

7. Denature the dsDNA gDNA fragments into ssDNA templates for cluster amplifications by *in situ* PCR.

B. *Sequencing*

1. Perform *in situ* PCR to amplify ssDNA templates into clusters.

2. Use NGS machine(s) for sequencing. This step produces a large amount of "raw reads" normally enough to make a 30-fold coverage of the target genome.

C. *Data processing*

Clean up contaminants, non-specific reads and low quality reads. This step generates a set of "quality reads", either paired-end (PE) or single end (SE), for assembly.

D. *Genome assembly*

1. Use the overlapped sequence regions among the quality reads as a guideline to perform *de novo* assembly. This step generates a set of contigs.

2. Use paired-end ditag/MP reads or other methods to identify the order of contigs. This step generates "scaffolds" which consist of ordered contigs and gaps with defined length, while some contigs remain un-ordered.

3. Narrow down gap size and reduce gap number. This can be done by various methods. For example, one can use the quality reads to map against the draft genome to extend the flanking regions of a gap. Alternatively, one also use PCR to fill out gaps. Sometimes, mRNA sequences from a transcriptome library can be

used as well, although the expressed sequences can cover only a limited portion.

Regarding gaps, here are some useful information. Assuming a Poisson distribution of sequence coverage,

1. $P_o = e^{-c}$, where P_o is the probability of unsequenced regions, c is the sequence coverage.
2. $L_g = Le^{-c}$, where L_g is the total gap length, L is the genome size, c is the sequence coverage.
3. The average gap size $= L_g/n$, where n is the number of gaps.

UCSC GENOME BROWSER

UCSC has a large collection of reference genomes and working draft assemblies, together with their associated information. In fact, UCSC does not produce genome assemblies, but instead obtains genome assemblies from NCBI, Sanger, or other standard resources. UCSC organizes genomic sequence data into the following order: group -> genome -> assembly -> position. There are five well-defined groups (*i.e.*, Mammal, Vertebrate, Deuterostome, Insect, and Nematode). Readers are recommended to login and play with the website for better understanding of its operation and the usage of the great resource of genome sequences and related information.

REFERENCES

Arabidopsis Genome, I. Arabidopsis Genome Initiative. (2000). Analysis of the genome sequence of the flowering plant Arabidopsis thaliana. *Nature, 408*(6814), 796-815.
[http://dx.doi.org/10.1038/35048692] [PMID: 11130711]

International Human Genome Sequencing Consortium. (2004). Finishing the euchromatic sequence of the human genome. *Nature,* Oct 21;*431*(7011), 931-945.
[http://dx.doi.org/10.1038/nature03001] [PMID: 15496913]

Lander, E.S., Linton, L.M., Birren, B., Nusbaum, C., Zody, M.C., Baldwin, J., Devon, K., Dewar, K., Doyle, M., FitzHugh, W., Funke, R., Gage, D., Harris, K., Heaford, A., Howland, J., Kann, L., Lehoczky, J., LeVine, R., McEwan, P., McKernan, K., Meldrim, J., Mesirov, J.P., Miranda, C., Morris, W., Naylor, J., Raymond, C., Rosetti, M., Santos, R., Sheridan, A., Sougnez, C., Stange-Thomann, N., Stojanovic, N., Subramanian, A., Wyman, D., Rogers, J., Sulston, J., Ainscough, R., Beck, S., Bentley, D., Burton, J., Clee, C., Carter, N., Coulson, A., Deadman, R., Deloukas, P., Dunham, A., Dunham, I., Durbin, R., French, L., Grafham, D., Gregory, S., Hubbard, T., Humphray, S., Hunt, A., Jones, M., Lloyd, C., McMurray, A., Matthews, L., Mercer, S., Milne, S., Mullikin, J.C., Mungall, A., Plumb, R., Ross, M., Shownkeen, R., Sims, S., Waterston, R.H., Wilson, R.K., Hillier, L.W., McPherson, J.D., Marra, M.A., Mardis, E.R., Fulton, L.A., Chinwalla, A.T., Pepin, K.H., Gish, W.R., Chissoe, S.L., Wendl, M.C., Delehaunty, K.D., Miner, T.L., Delehaunty, A., Kramer, J.B., Cook, L.L., Fulton, R.S., Johnson, D.L., Minx, P.J., Clifton, S.W., Hawkins, T., Branscomb, E., Predki, P., Richardson, P., Wenning, S., Slezak, T., Doggett, N., Cheng, J.F., Olsen, A., Lucas, S., Elkin, C., Uberbacher, E., Frazier, M., Gibbs, R.A., Muzny, D.M., Scherer, S.E., Bouck, J.B., Sodergren, E.J., Worley, K.C., Rives, C.M., Gorrell, J.H., Metzker, M.L., Naylor, S.L., Kucherlapati, R.S., Nelson, D.L., Weinstock, G.M., Sakaki, Y., Fujiyama, A., Hattori, M., Yada, T., Toyoda, A., Itoh, T., Kawagoe, C., Watanabe, H., Totoki, Y., Taylor, T., Weissenbach, J., Heilig, R., Saurin, W., Artiguenave, F., Brottier, P., Bruls, T., Pelletier, E., Robert, C., Wincker, P., Smith, D.R., Doucette-Stamm, L., Rubenfield, M., Weinstock, K., Lee,

H.M., Dubois, J., Rosenthal, A., Platzer, M., Nyakatura, G., Taudien, S., Rump, A., Yang, H., Yu, J., Wang, J., Huang, G., Gu, J., Hood, L., Rowen, L., Madan, A., Qin, S., Davis, R.W., Federspiel, N.A., Abola, A.P., Proctor, M.J., Myers, R.M., Schmutz, J., Dickson, M., Grimwood, J., Cox, D.R., Olson, M.V., Kaul, R., Raymond, C., Shimizu, N., Kawasaki, K., Minoshima, S., Evans, G.A., Athanasiou, M., Schultz, R., Roe, B.A., Chen, F., Pan, H., Ramser, J., Lehrach, H., Reinhardt, R., McCombie, W.R., de la Bastide, M., Dedhia, N., Blöcker, H., Hornischer, K., Nordsiek, G., Agarwala, R., Aravind, L., Bailey, J.A., Bateman, A., Batzoglou, S., Birney, E., Bork, P., Brown, D.G., Burge, C.B., Cerutti, L., Chen, H.C., Church, D., Clamp, M., Copley, R.R., Doerks, T., Eddy, S.R., Eichler, E.E., Furey, T.S., Galagan, J., Gilbert, J.G., Harmon, C., Hayashizaki, Y., Haussler, D., Hermjakob, H., Hokamp, K., Jang, W., Johnson, L.S., Jones, T.A., Kasif, S., Kaspryzk, A., Kennedy, S., Kent, W.J., Kitts, P., Koonin, E.V., Korf, I., Kulp, D., Lancet, D., Lowe, T.M., McLysaght, A., Mikkelsen, T., Moran, J.V., Mulder, N., Pollara, V.J., Ponting, C.P., Schuler, G., Schultz, J., Slater, G., Smit, A.F., Stupka, E., Szustakowski, J., Thierry-Mieg, D., Thierry-Mieg, J., Wagner, L., Wallis, J., Wheeler, R., Williams, A., Wolf, Y.I., Wolfe, K.H., Yang, S.P., Yeh, R.F., Collins, F., Guyer, M.S., Peterson, J., Felsenfeld, A., Wetterstrand, K.A., Patrinos, A., Morgan, M.J., de Jong, P., Catanese, J.J., Osoegawa, K., Shizuya, H., Choi, S., Chen, Y.J. International Human Genome Sequencing Consortium. (2001). Initial sequencing and analysis of the human genome. *Nature, 409*(6822), 860-921. [http://dx.doi.org/10.1038/35057062] [PMID: 11237011]

Waterston, R.H., Lindblad-Toh, K., Birney, E., Rogers, J., Abril, J.F., Agarwal, P., Agarwala, R., Ainscough, R., Alexandersson, M., An, P., Antonarakis, S.E., Attwood, J., Baertsch, R., Bailey, J., Barlow, K., Beck, S., Berry, E., Birren, B., Bloom, T., Bork, P., Botcherby, M., Bray, N., Brent, M.R., Brown, D.G., Brown, S.D., Bult, C., Burton, J., Butler, J., Campbell, R.D., Carninci, P., Cawley, S., Chiaromonte, F., Chinwalla, A.T., Church, D.M., Clamp, M., Clee, C., Collins, F.S., Cook, L.L., Copley, R.R., Coulson, A., Couronne, O., Cuff, J., Curwen, V., Cutts, T., Daly, M., David, R., Davies, J., Delehaunty, K.D., Deri, J., Dermitzakis, E.T., Dewey, C., Dickens, N.J., Diekhans, M., Dodge, S., Dubchak, I., Dunn, D.M., Eddy, S.R., Elnitski, L., Emes, R.D., Eswara, P., Eyras, E., Felsenfeld, A., Fewell, G.A., Flicek, P., Foley, K., Frankel, W.N., Fulton, L.A., Fulton, R.S., Furey, T.S., Gage, D., Gibbs, R.A., Glusman, G., Gnerre, S., Goldman, N., Goodstadt, L., Grafham, D., Graves, T.A., Green, E.D., Gregory, S., Guigó, R., Guyer, M., Hardison, R.C., Haussler, D., Hayashizaki, Y., Hillier, L.W., Hinrichs, A., Hlavina, W., Holzer, T., Hsu, F., Hua, A., Hubbard, T., Hunt, A., Jackson, I., Jaffe, D.B., Johnson, L.S., Jones, M., Jones, T.A., Joy, A., Kamal, M., Karlsson, E.K., Karolchik, D., Kasprzyk, A., Kawai, J., Keibler, E., Kells, C., Kent, W.J., Kirby, A., Kolbe, D.L., Korf, I., Kucherlapati, R.S., Kulbokas, E.J., Kulp, D., Landers, T., Leger, J.P., Leonard, S., Letunic, I., Levine, R., Li, J., Li, M., Lloyd, C., Lucas, S., Ma, B., Maglott, D.R., Mardis, E.R., Matthews, L., Mauceli, E., Mayer, J.H., McCarthy, M., McCombie, W.R., McLaren, S., McLay, K., McPherson, J.D., Meldrim, J., Meredith, B., Mesirov, J.P., Miller, W., Miner, T.L., Mongin, E., Montgomery, K.T., Morgan, M., Mott, R., Mullikin, J.C., Muzny, D.M., Nash, W.E., Nelson, J.O., Nhan, M.N., Nicol, R., Ning, Z., Nusbaum, C., O'Connor, M.J., Okazaki, Y., Oliver, K., Overton-Larty, E., Pachter, L., Parra, G., Pepin, K.H., Peterson, J., Pevzner, P., Plumb, R., Pohl, C.S., Poliakov, A., Ponce, T.C., Ponting, C.P., Potter, S., Quail, M., Reymond, A., Roe, B.A., Roskin, K.M., Rubin, E.M., Rust, A.G., Santos, R., Sapojnikov, V., Schultz, B., Schultz, J., Schwartz, M.S., Schwartz, S., Scott, C., Seaman, S., Searle, S., Sharpe, T., Sheridan, A., Shownkeen, R., Sims, S., Singer, J.B., Slater, G., Smit, A., Smith, D.R., Spencer, B., Stabenau, A., Stange-Thomann, N., Sugnet, C., Suyama, M., Tesler, G., Thompson, J., Torrents, D., Trevaskis, E., Tromp, J., Ucla, C., Ureta-Vidal, A., Vinson, J.P., Von Niederhausern, A.C., Wade, C.M., Wall, M., Weber, R.J., Weiss, R.B., Wendl, M.C., West, A.P., Wetterstrand, K., Wheeler, R., Whelan, S., Wierzbowski, J., Willey, D., Williams, S., Wilson, R.K., Winter, E., Worley, K.C., Wyman, D., Yang, S., Yang, S.P., Zdobnov, E.M., Zody, M.C., Lander, E.S. Mouse Genome Sequencing Consortium. (2002). Initial sequencing and comparative analysis of the mouse genome. *Nature, 420*(6915), 520-562. [http://dx.doi.org/10.1038/nature01262] [PMID: 12466850]

Myers, E.W., Sutton, G.G., Delcher, A.L., Dew, I.M., Fasulo, D.P., Flanigan, M.J., Kravitz, S.A., Mobarry, C.M., Reinert, K.H., Remington, K.A., Anson, E.L., Bolanos, R.A., Chou, H.H., Jordan, C.M., Halpern, A.L.,

Lonardi, S., Beasley, E.M., Brandon, R.C., Chen, L., Dunn, P.J., Lai, Z., Liang, Y., Nusskern, D.R., Zhan, M., Zhang, Q., Zheng, X., Rubin, G.M., Adams, M.D., Venter, J.C. (2000). A whole-genome assembly of Drosophila. *Science, 287*(5461), 2196-2204.
[http://dx.doi.org/10.1126/science.287.5461.2196] [PMID: 10731133]

Schmutz, J., Wheeler, J., Grimwood, J., Dickson, M., Yang, J., Caoile, C., Bajorek, E., Black, S., Chan, Y.M., Denys, M., Escobar, J., Flowers, D., Fotopulos, D., Garcia, C., Gomez, M., Gonzales, E., Haydu, L., Lopez, F., Ramirez, L., Retterer, J., Rodriguez, A., Rogers, S., Salazar, A., Tsai, M., Myers, R.M. (2004). Quality assessment of the human genome sequence. *Nature, 429*(6990), 365-368.
[http://dx.doi.org/10.1038/nature02390] [PMID: 15164052]

Venter, J.C., Adams, M.D., Myers, E.W., Li, P.W., Mural, R.J., Sutton, G.G., Smith, H.O., Yandell, M., Evans, C.A., Holt, R.A., Gocayne, J.D., Amanatides, P., Ballew, R.M., Huson, D.H., Wortman, J.R., Zhang, Q., Kodira, C.D., Zheng, X.H., Chen, L., Skupski, M., Subramanian, G., Thomas, P.D., Zhang, J., Gabor Miklos, G.L., Nelson, C., Broder, S., Clark, A.G., Nadeau, J., McKusick, V.A., Zinder, N., Levine, A.J., Roberts, R.J., Simon, M., Slayman, C., Hunkapiller, M., Bolanos, R., Delcher, A., Dew, I., Fasulo, D., Flanigan, M., Florea, L., Halpern, A., Hannenhalli, S., Kravitz, S., Levy, S., Mobarry, C., Reinert, K., Remington, K., Abu-Threideh, J., Beasley, E., Biddick, K., Bonazzi, V., Brandon, R., Cargill, M., Chandramouliswaran, I., Charlab, R., Chaturvedi, K., Deng, Z., Di Francesco, V., Dunn, P., Eilbeck, K., Evangelista, C., Gabrielian, A.E., Gan, W., Ge, W., Gong, F., Gu, Z., Guan, P., Heiman, T.J., Higgins, M.E., Ji, R.R., Ke, Z., Ketchum, K.A., Lai, Z., Lei, Y., Li, Z., Li, J., Liang, Y., Lin, X., Lu, F., Merkulov, G.V., Milshina, N., Moore, H.M., Naik, A.K., Narayan, V.A., Neelam, B., Nusskern, D., Rusch, D.B., Salzberg, S., Shao, W., Shue, B., Sun, J., Wang, Z., Wang, A., Wang, X., Wang, J., Wei, M., Wides, R., Xiao, C., Yan, C., Yao, A., Ye, J., Zhan, M., Zhang, W., Zhang, H., Zhao, Q., Zheng, L., Zhong, F., Zhong, W., Zhu, S., Zhao, S., Gilbert, D., Baumhueter, S., Spier, G., Carter, C., Cravchik, A., Woodage, T., Ali, F., An, H., Awe, A., Baldwin, D., Baden, H., Barnstead, M., Barrow, I., Beeson, K., Busam, D., Carver, A., Center, A., Cheng, M.L., Curry, L., Danaher, S., Davenport, L., Desilets, R., Dietz, S., Dodson, K., Doup, L., Ferriera, S., Garg, N., Gluecksmann, A., Hart, B., Haynes, J., Haynes, C., Heiner, C., Hladun, S., Hostin, D., Houck, J., Howland, T., Ibegwam, C., Johnson, J., Kalush, F., Kline, L., Koduru, S., Love, A., Mann, F., May, D., McCawley, S., McIntosh, T., McMullen, I., Moy, M., Moy, L., Murphy, B., Nelson, K., Pfannkoch, C., Pratts, E., Puri, V., Qureshi, H., Reardon, M., Rodriguez, R., Rogers, Y.H., Romblad, D., Ruhfel, B., Scott, R., Sitter, C., Smallwood, M., Stewart, E., Strong, R., Suh, E., Thomas, R., Tint, N.N., Tse, S., Vech, C., Wang, G., Wetter, J., Williams, S., Williams, M., Windsor, S., Winn-Deen, E., Wolfe, K., Zaveri, J., Zaveri, K., Abril, J.F., Guigó, R., Campbell, M.J., Sjolander, K.V., Karlak, B., Kejariwal, A., Mi, H., Lazareva, B., Hatton, T., Narechania, A., Diemer, K., Muruganujan, A., Guo, N., Sato, S., Bafna, V., Istrail, S., Lippert, R., Schwartz, R., Walenz, B., Yooseph, S., Allen, D., Basu, A., Baxendale, J., Blick, L., Caminha, M., Carnes-Stine, J., Caulk, P., Chiang, Y.H., Coyne, M., Dahlke, C., Mays, A., Dombroski, M., Donnelly, M., Ely, D., Esparham, S., Fosler, C., Gire, H., Glanowski, S., Glasser, K., Glodek, A., Gorokhov, M., Graham, K., Gropman, B., Harris, M., Heil, J., Henderson, S., Hoover, J., Jennings, D., Jordan, C., Jordan, J., Kasha, J., Kagan, L., Kraft, C., Levitsky, A., Lewis, M., Liu, X., Lopez, J., Ma, D., Majoros, W., McDaniel, J., Murphy, S., Newman, M., Nguyen, T., Nguyen, N., Nodell, M., Pan, S., Peck, J., Peterson, M., Rowe, W., Sanders, R., Scott, J., Simpson, M., Smith, T., Sprague, A., Stockwell, T., Turner, R., Venter, E., Wang, M., Wen, M., Wu, D., Wu, M., Xia, A., Zandieh, A., Zhu, X (2001). The sequence of the human genome. *Science, 291*(5507), 1304-1351.
[http://dx.doi.org/10.1126/science.1058040] [PMID: 11181995]

CHAPTER 9

Exome Sequencing: Genome Sequencing Focusing on Exonic Regions

Abstract: While whole genome sequencing (WGS) remains costly and requires intensive labor and elaborate analytical tools for assembly, whole exome sequencing (WES) is relatively cheaper and easier. Compared to WGS, WES can be considered as an efficient approach when the protein-coding regions are the only concern, because this type of sequencing focuses on the exon regions and its desired sequencing depth can be easily reached. WES is frequently confused with transcriptome analysis because both types of libraries contain solely the exonal sequences. However, the former is generated from genomic DNA fragments, while the latter from expressed mRNA molecules. Readers are asked to distinguish the differences between these two libraries beforehand.

Keywords: Whole exome sequencing, WES, Whole genome sequencing, WGS.

Definition of Terminologies
Exome

Exome is a scientific term representing the collection of exonal sequences. Thus, exome is part of the genome, and exomics is part of the genomics.

INTRODUCTION

There are ~180,000 protein-coding *exons* in the human genome. These exons occupy approximately 1% of the human genome and yet harbor about 85% disease-causing "genetic" mutations, which can be directly detected by next-gen sequencing (Gilissen *et al.*, 2012). Whole exome sequencing (WES) application was first reported in 2009 (Choi *et al.*, 2009; Ng *et al.*, 2009). By combining exome capture, or (exome) target enrichment, approaches with NGS technologies, WES has become a robust approach for the identification sequence variations responsible for common and rare Mendelian diseases. WES enhances the study of SNPs (single nucleotide polymorphisms, or single nucleotide variations (SNVs)) and indels (insertions and deletions) between diseased and normal tissues and leads to the identification of disease genes.

Whole exome sequencing heavily relies on specific selection of exonic DNA from fragmented genomic DNA preparations (of an individual) (see figure shown below). There are a number of exome capture approaches: multiplex PCR which uses multiple pairs of PCR primers in a single polymerase chain reaction to amplify multiple exons from a genomic DNA preparation, molecular inversion probe (MIP), microarray hybridization capture on glass slides as pioneered by Roche NimbleGen (Sequence Capture Human Exome 2.1M Arrays), and in-solution capture on beads adopted by Illumina (TruSeq Kit), and also by Roche NimbleGen (SeqCap EZ Exome Library Kit) (Mamanova *et al.*, 2010). These exome capture methods bypass the intronic regions, and frequently the 5' and 3' untranslated regions (UTRs) as well, to selectively enrich the exonic DNA for next-generation sequencing (Fig. 1).

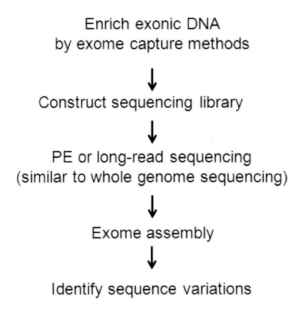

Fig. (1). Experimental procedure of exome sequencing.

SEQUENCING LIBRARY CONSTRUCTION

Procedure for preparing an exome sequencing library remains very similar to that for other types of sequencing, except that an exome capture method has to be decided for initial exonic DNA preparation.

As mentioned above, there are a number of target enrichment kits made commercially available by Roche NimbleGen (using microarray for hybridization capture and in-solution sequence capture), Illumina, and Agilent, *etc*. These kits made the sequencing library construction for exome sequencing simpler and straightforward.

Multiplex PCR is not feasible for whole exome sequencing. It is only suitable for the amplification of limited number of exons, because bias increases along with the number of exons under amplification due to potential imbalanced amplification efficiency between exons. PCR bias can result from the difference in length, primer specificity, enzymatic activity, the amount of dNTPs provided, *etc*. However, multiplex PCR method can still be considered when limited number of exons are to be examined.

Exome target enrichment strategies for NGS was described by Mamanova *et al.*, (Mamanova *et al.*, 2010).

SEQUENCE DATA ANALYSIS

The first step in sequence data processing is to remove (discard) unwanted reads from raw data. The unwanted reads include low-quality reads, contaminated sequences such as sequences partially or fully occupied by vector or adaptor sequences, repetitive sequences (*e.g.* polyA$_9$/T$_9$/G$_9$/C$_9$ and sequences with recognizable repetitive patterns which may cause these sequences to map to numerous genomic locations), sequences from rRNAs or tRNAs when these species are not under investigation, and questionable reads (*e.g.* intronic sequences when exome sequencing is the research focus).

Since the sequenced DNA fragments are supposed to be exon-originated while certain level of uncertainty remains, we need to calculate the percentage of reads mapped to the exonic regions over the total reads. Considering the possibility of non-specific capture, 100% is not possible, but normally over 80% is expected. Sequences mapped to non-exonic regions are normally discarded from sequence pool. Sequences mapped to exonic regions are aligned with software to identify the genetic mutations. For current status of NGS capability, at least 20-fold coverage is regularly desired.

Processing and analysis of WES data for the identification of novel gene mutations associated with rare Mendelian diseases was well illustrated in the review article by Gilissen and colleagues (Gilissen *et al.*, 2012). As described by the authors, the number of variant calls can vary significantly depending on the capture method and the sequencing platform been used. Roughly the initial number of variant calls can reach tens of thousands per case. Through the routine procedure of sequence quality control to remove the low quality sequences followed by filtering out the ones located in the non-coding regions, most false positive calls can be removed, causing the number drop down to about a few thousands. This number can be further reduced by removing know variants, making the number of variant calls drops for another 10-folds. Now, working on a few hundred genes seems to be much easier comparing to the initial number, but the reduction itself requires fine-tuning using certain case-dependent strategies, including linkage strategy, homozygosity strategy, double-hit strategy, overlap strategy, *de novo* strategy, and candidate strategy, *etc.* (Gilissen *et al.*, 2012).

SINGLE NUCLEOTIDE POLYMORPHISM

Single nucleotide polymorphism (SNP, pronounced *snip*) is a sequence variation which may occur naturally between individuals of the same species. SNPs are found not only in the intergenic and intronic regions, but also in the exonic regions. These genetic variations have been used as DNA fingerprinting to distinguish individuals, and some SNPs are associated with potential risks of developing a disease.

REFERENCES

Choi, M., Scholl, U.I., Ji, W., Liu, T., Tikhonova, I.R., Zumbo, P., Nayir, A., Bakkaloğlu, A., Ozen, S., Sanjad, S., Nelson-Williams, C., Farhi, A., Mane, S., Lifton, R.P. (2009). Genetic diagnosis by whole exome capture and massively parallel DNA sequencing. *Proc. Natl. Acad. Sci. USA, 106*(45), 19096-19101. [http://dx.doi.org/10.1073/pnas.0910672106] [PMID: 19861545]

Gilissen, C., Hoischen, A., Brunner, H.G., Veltman, J.A. (2012). Disease gene identification strategies for exome sequencing. *Eur. J. Hum. Genet., 20*(5), 490-497. [http://dx.doi.org/10.1038/ejhg.2011.258] [PMID: 22258526]

Mamanova, L., Coffey, A.J., Scott, C.E., Kozarewa, I., Turner, E.H., Kumar, A., Howard, E., Shendure, J., Turner, D.J. (2010). Target-enrichment strategies for next-generation sequencing. *Nat. Methods, 7*(2), 111-118. [http://dx.doi.org/10.1038/nmeth.1419] [PMID: 20111037]

Ng, S.B., Turner, E.H., Robertson, P.D., Flygare, S.D., Bigham, A.W., Lee, C., Shaffer, T., Wong, M., Bhattacharjee, A., Eichler, E.E., Bamshad, M., Nickerson, D.A., Shendure, J. (2009). Targeted capture and massively parallel sequencing of 12 human exomes. *Nature, 461*(7261), 272-276. [http://dx.doi.org/10.1038/nature08250] [PMID: 19684571]

CHAPTER 10

Transcriptome Analysis

Abstract: Transcriptome analysis, or transcriptome sequencing, concerns the transcript sequences transcribed from the genome of a specific cell type at specific time and growth conditions. Previous studies have clearly demonstrated that, besides messenger RNA (mRNA), the transcribed RNA sequences also contain large amounts of ribosomal RNA (rRNA), transfer RNA (tRNA) and small-sized non-coding RNA (ncRNA). Transcriptome analysis focuses mainly on mRNA, and sometimes, certain types of ncRNA species which may be co-isolated with mRNA when gene expression and regulation are of the major concern. From transcriptome sequencing, a number of biological information can be retrieved. These include gene expression level, transcriptome landscape across the entire genome, Gene Ontology, pathway, *etc*. Notice that transcriptome analysis normally refers to the whole transcriiptome analysis of a cell population. The result is in fact a combination of millions of potentially diversified single-cell transcriptomes.

Keywords: Gene Ontology, GO, KEGG, Pathway, RPKM, Transcriptome.

Definition of Terminologies

Ttranscriptome

The RNA products of gene expression from a cell or a cell population at a particular status and time point constitute a "transcriptome". In a broad sense, a transcriptome should include mRNAs to be translated into a "proteome", rRNAs to be used for making ribosomes for protein synthesis, tRNAs to be used in carrying amino acids for protein synthesis, and many kinds of small-sized non-coding RNAs, including so-called small RNAs (sRNAs, ~50-250 nucleotides), microRNAs (miRNAs, ~ 17-25 nucleotides), and Piwi-interacting RNA (piRNA, typically 24-32 nucleotides). Since the roles of rRNAs and tRNAs have been well-studied and defined, most current transcriptomic investigations focus on mRNA transcriptomes and non-coding small-sized RNA molecules, especially sRNAs, miRNAs, and piRNAs, and the like.

INTRODUCTION

There are about 24,000 protein-coding genes in the human genome. In a differentiated cell population, only about half of these human genes are expressed, or transcribed, into mRNAs, while single cells express even fewer genes. The expression of protein-coding genes is accompanied by the expression of a few thousand non-coding genes which selectively degrade or inhibit the translational efficiency of their target mRNAs in response to the environmental cues. The expression of protein-coding and non-coding genes are fine-tuned in a coordinated fashion to balance physiological conditions in a homeostatic manner.

Gene expression, or transcription, involves multiple steps of hierarchical regulation. During cell differentiation, genes, and frequently, their corresponding regulatory elements, are packaged in various manners based on their prospects of future usage. Those packaged in heterochromatin domains are likely to become unusable; those maintained in the facultative heterochromatin domains may still be likely to be used; and those in the euchromatin domains will be translated. Notice that the arrangement of heterochromatin, euchromatin, and facultative euchromatin is tissue type-dependent because different tissues use different sets of genes. Compact packaging in the heterochromatin domains makes the encompassed genes and their corresponding regulatory elements inaccessible by transcription factors (TFs), cofactors, or transcriptional machinery. Epigenetic modifications further separate the euchromatin domains, and thus their encompassed genes, into various states of readiness for transcription. Extracellular molecular signals (*e.g.*, hormones from the endocrine system, cytokines from the immune system, and extracellular ligands) regulate intracellular states and activate corresponding TFs and their cofactors. Activated TFs recruit their cofactors, including histone modifiers such as HDAC and methyltransferase, and bind to their target motifs in the euchromatin. These events result in limited alteration in epigenetic modifications, leading to a transcriptional activation and/or repression of certain sets of responsive genes.Taken together all these molecular events taking place in the microscopic world, one can understand that transcriptional activation and repression are highly regulated.

IMPACT OF GENOME ASSEMBLY ON TRANSCRIPTOME ANALYSIS

Before genome assemblies were made available during the turn of the twenty-first century, transcriptome analysis was heavily relying on microarray, SAGE (serial analysis of gene expression), and EST (expressed sequence tag), especially microarray. Once being a well commercialized and well-established technology, microarray extracts gene-specific sequences from mRNA to make thousands or tens of thousands of probes. With the assistance of instruments, some of which are semi-automated, probes are hybridized to the transcripts present in transcriptome libraries. After a few washes to remove nonspecific hybridization, signal intensity of a hybridized spot is subtracted by the signal instensity of the nearest control (background). By so doing, the level of expressed transcript is quantified as the hybridization intensity, most likely to be a non-integer value. On the other hand, both SAGE and EST rely on sequencing to profile transcriptomes and virtual databases built from known gene-specific sequences to correlate the expressed transcripts with their corresponding genes. The expression levels can be presented as integers.

It should be noted that originally these technologies could only detect the expression of known genes while novel genes were ignored. This was because the design of microarray probes relied on known genetic sequences and the setup of the SAGE virtual database also relied on known sequences. Genome assemblies now provide complete sets of genomic sequence information, including known genes and previously unknown (novel) sequences for us to compare with our local transcriptome data. Accordingly, experimental and analytical strategies have to be changed to integrate genome assembly information in order to enhance transcriptome analysis.

This chapter focuses on mRNA transcriptome of protein-coding genes. The study of gene expression of a cell population, tissue, organ, or an organism relies on whole transcriptome sequencing. Whole transcriptome sequencing allows us to discover transcriptional (gene expression) fluctuations such as the upregulated and downregulated genes, grouping of the fluctuated genes by Gene Ontology, and biological pathway analysis (Carninci *et al.*, 2005; Chiu *et al.*, 2007; Consortium *et al.*, 2007; Ng ., 2005).

CONSTRUCTION OF TRANSCRIPTOMIC SEQUENCING LIBRARIES

For the construction of a whole transcriptome library, it is essential to isolate total RNA from cells, followed by isolation of mRNA from the total RNA and reverse transcription using oligo-dT to generate a cDNA library from mRNAs. To achieve optimal sequence specificity and sequencing efficiency, fragmentation of target molecules is essential and can be done on mRNA or cDNA.

We strongly recommend mRNA fragmentation for a number of reasons, in particular, the following. First, the efficiency of reverse transcription of mRNA molecules into cDNA molecules is size-dependent; long mRNA molecules require more time to complete the process. Thus, this process favors short mRNAs and is likely to introduce bias. Second, if there is degradation in the isolated mRNA, only the regions containing poly-As can be converted into cDNA by reverse transcription. This potential problem will create another layer of bias leading to a sequence-to-genome mapping mistakenly concentrated on the last few exons. To minimize these potential sources of bias, mRNA fragmentation is advised.

PAIRED-END DITAG SEQUENCING *VS.* SHOTGUN FRAGMENT SEQUENCING OF TRANSCRIPTOME LIBRARIES

As mentioned previously, a transcriptome library can be built into either a shotgun fragment (SF) sequencing library or a paired-end ditag (PED) sequencing library (Fig. 1). The former will be subjected to either fragment sequencing (using a single (set) of sequencing primer(s)) or paired-end (PE) sequencing (using forward and reverse sequencing primers), while the latter is solely for paired-end ditag sequencing.

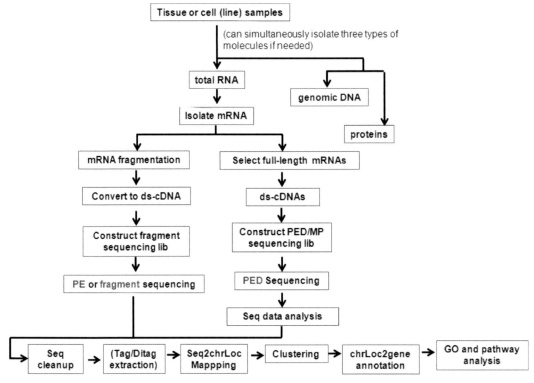

Fig. (1). Preparation of sequencing libraries for whole transcriptome analysis.

SEQUENCE DATA PROCESSING

Overe the past few years, a number of analytical pipelines/software tools have been made commercially available. These pipelines and software tools, including Galaxy, Bowtie sequence aligner, CLC and those provided by sequencer manufacturers, are able to selectively process SF libraries, PE libraries, and PED libraries in a customized fashion.

Generally speaking, there are steps every software has to take in order to produce reliable results from piles of sequences. These essential steps include sequence cleanup (*i.e.*, selection of quality sequences by removing low quality reads, vector sequences, contaminant sequences, *etc.*), sequence extraction (*i.e.*, extraction of desired sequences), reorganization of quality reads, and sequence alignment/mapping against a reference genome. A careful bioinformatician may also want to upload raw data and processed data into a database (*e.g.*, mySQL or Oracle) to facilitate data management and tracking. Unfortunately, today many analytical pipelines and software tools tend to skip this part, making data tracking very difficult.

CALCULATION OF GENE EXPRESSION LEVEL

Calculation of transcriptional levels for all expressed genes is important. Through cross-library comparison upon gene expression levels, one can identify upregulated and downregulated genes across the libraries under comparison. Further analysis using pathway databases, the most significantly upregulated and most significantly downregulated biological pathways can then be identified.

I. *Counts (or copy number) per million*
   - Suitable, and has been used, for paired-end ditag or miRNA libraries
II. *RPKM (Reads per kilo base per million)*
   - Suitable, and has been commonly used, for shotgun RNA-Seq libraries

DISPLAY OF TRANSCRIPTOMIC SEQUENCE DATA

To facilitate library comparison, data sharing and communication, sequences and related information should be displayed by software in a descriptive and easy-to-understand manner. Here, we use two examples to demonstrate how a library data can be displayed in a computer terminal.

Fig. (2) shows a local paired-end ditag library's data (green tracks on top) displayed side-by-side with a reference genome retrieved from the UCSC genome browser. Notice that each paired-end ditag track is composed of a 5' tag, a 3' tag, and a central uncertain region. We know only the terminal tag sequences, while the sequence in between remains unknown.

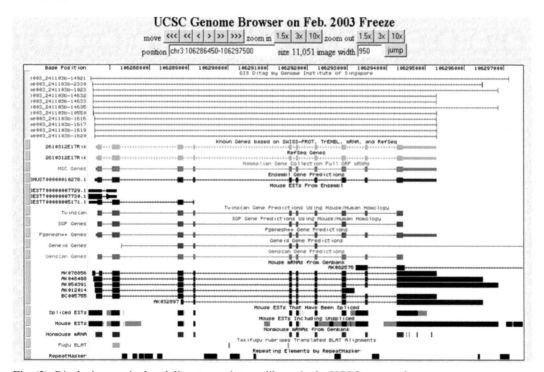

Fig. (2). Displaying a paired-end ditag transcriptome library in the UCSC genome browser.

Fig. (**3**) shows the results of multiple single-cell transcriptome libraries produced and analyzed by the RNA-Seq approach. Notice that sequence reads are mapped to the exonic regions. However, connections between exons can be obscure. Software designed to display multiple libraries simultaneously enhances cross-library comparisons.

Cross-library comparison of expression profiles of single-cell transcriptome libraries

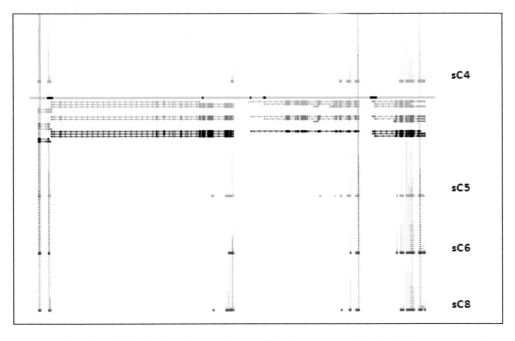

Fig. (3). Display of multiple single-cell transcriptome libraries generated by the RNA-seq approach using GeneWalk software made in-house. Local single cell transcriptome libraries, labeled sC4 (green), sC5 (orange), sC6 (blue), and sC8 (red), are displayed side-by-side with the chromosome (orange bar right below sC4) and gene tracks (shown right below the chromosome track) downloaded from the UCSC genome browser.

CATEGORIZATION OF UPREGULATED AND DOWNREGULATED GENES BY GENE ONTOLOGY ANALYSIS

As described by *Wikipedia*: "Gene ontology, or GO, is a major bioinformatics initiative to unify the representation of gene and gene product attributes across all species. More specifically, the project aims to: 1) Maintain and develop its controlled vocabulary of gene and gene product attributes; 2) Annotate genes and gene products, and assimilate and disseminate annotation data; 3) Provide tools for easy access to all aspects of the data provided by the project, and to enable functional interpretation of experimental data using the GO,...." GO is made of three layers: cellular component, molecular function, and biological process. GO analysis on upregulated and downregulated genes helps researchers understand which genes are involved in the most significant affected parts/functions of the cells under investigation.

IDENTIFICATION OF MOST SIGNIFICANTLY UPREGULATED / DOWNREGULATED PATHWAYS

Pathways represent the functional units that define the overall function of a cell. Transcriptome analysis at the gene level, or GO analysis alone, is not sufficient to compose a comprehensive story about intracellular conditions. Pathway analysis is able to help us gain valuable insight into the alterations in cellular functions.

There are a number of pathway databases available online. These include the David pathway viewer, BioCarta pathway database, Kyoto Encyclopedia of Genes and Genomes (KEGG) pathway database, and many others. Personally, I prefer to use KEGG database for pathway analysis because of its flexibility and user friendliness (Chiu *et al.*, 2007). However, interested readers are recommended to review at least a subset of these databases and to select one or more that will meet the particular needs of their studies.

REFERENCES

Carninci, P., Kasukawa, T., Katayama, S., Gough, J., Frith, M.C., Maeda, N., Oyama, R., Ravasi, T., Lenhard, B., Wells, C., Kodzius, R., Shimokawa, K., Bajic, V.B., Brenner, S.E., Batalov, S., Forrest, A.R., Zavolan, M., Davis, M.J., Wilming, L.G., Aidinis, V., Allen, J.E., Ambesi-Impiombato, A., Apweiler, R., Aturaliya, R.N., Bailey, T.L., Bansal, M., Baxter, L., Beisel, K.W., Bersano, T., Bono, H., Chalk, A.M., Chiu, K.P., Choudhary, V., Christoffels, A., Clutterbuck, D.R., Crowe, M.L., Dalla, E., Dalrymple, B.P., de Bono, B., Della Gatta, G., di Bernardo, D., Down, T., Engstrom, P., Fagiolini, M., Faulkner, G., Fletcher, C.F., Fukushima, T., Furuno, M., Futaki, S., Gariboldi, M., Georgii-Hemming, P., Gingeras, T.R., Gojobori, T., Green, R.E., Gustincich, S., Harbers, M., Hayashi, Y., Hensch, T.K., Hirokawa, N., Hill, D., Huminiecki, L., Iacono, M., Ikeo, K., Iwama, A., Ishikawa, T., Jakt, M., Kanapin, A., Katoh, M., Kawasawa, Y., Kelso, J., Kitamura, H., Kitano, H., Kollias, G., Krishnan, S.P., Kruger, A., Kummerfeld, S.K., Kurochkin, I.V., Lareau, L.F., Lazarevic, D., Lipovich, L., Liu, J., Liuni, S., McWilliam, S., Madan Babu, M., Madera, M., Marchionni, L., Matsuda, H., Matsuzawa, S., Miki, H., Mignone, F., Miyake, S., Morris, K., Mottagui-Tabar, S., Mulder, N., Nakano, N., Nakauchi, H., Ng, P., Nilsson, R., Nishiguchi, S., Nishikawa, S., Nori, F., Ohara, O., Okazaki, Y., Orlando, V., Pang, K.C., Pavan, W.J., Pavesi, G., Pesole, G., Petrovsky, N., Piazza, S., Reed, J., Reid, J.F., Ring, B.Z., Ringwald, M., Rost, B., Ruan, Y., Salzberg, S.L., Sandelin, A., Schneider, C., Schönbach, C., Sekiguchi, K., Semple, C.A., Seno, S., Sessa, L., Sheng, Y., Shibata, Y., Shimada, H., Shimada, K., Silva, D., Sinclair, B., Sperling, S., Stupka, E., Sugiura, K., Sultana, R., Takenaka, Y., Taki, K., Tammoja, K., Tan, S.L., Tang, S., Taylor, M.S., Tegner, J., Teichmann, S.A., Ueda, H.R., van Nimwegen, E., Verardo, R., Wei, C.L., Yagi, K., Yamanishi, H., Zabarovsky, E., Zhu, S., Zimmer, A., Hide, W., Bult, C., Grimmond, S.M., Teasdale, R.D., Liu, E.T., Brusic, V., Quackenbush, J., Wahlestedt, C., Mattick, J.S., Hume, D.A., Kai, C., Sasaki, D., Tomaru, Y., Fukuda, S., Kanamori-Katayama, M., Suzuki, M., Aoki, J., Arakawa, T., Iida, J., Imamura, K., Itoh, M., Kato, T., Kawaji, H., Kawagashira, N., Kawashima, T., Kojima, M., Kondo, S., Konno, H., Nakano, K., Ninomiya, N., Nishio, T., Okada, M., Plessy, C., Shibata, K., Shiraki, T., Suzuki, S., Tagami, M., Waki, K., Watahiki, A., Okamura-Oho, Y., Suzuki, H., Kawai, J., Hayashizaki, Y. FANTOM Consortium; (2005). RIKEN Genome Exploration Research Group and Genome Science Group (Genome Network Project Core Group). (2005). The transcriptional landscape of the mammalian genome. *Science,* *309*(5740), 1559-1563.
[http://dx.doi.org/10.1126/science.1112014] [PMID: 16141072]

Chiu, K.P., Ariyaratne, P., Xu, H., Tan, A., Ng, P., Liu, E.T., Ruan, Y., Wei, C.L., Sung, W.K. (2007). Pathway aberrations of murine melanoma cells observed in Paired-End diTag transcriptomes. *BMC Cancer, 7*, 109.
[http://dx.doi.org/10.1186/1471-2407-7-109] [PMID: 17594473]

Birney, E., Stamatoyannopoulos, J.A., Dutta, A., Guigó, R., Gingeras, T.R., Margulies, E.H., Weng, Z., Snyder, M., Dermitzakis, E.T., Thurman, R.E., Kuehn, M.S., Taylor, C.M., Neph, S., Koch, C.M., Asthana, S., Malhotra, A., Adzhubei, I., Greenbaum, J.A., Andrews, R.M., Flicek, P., Boyle, P.J., Cao, H., Carter, N.P., Clelland, G.K., Davis, S., Day, N., Dhami, P., Dillon, S.C., Dorschner, M.O., Ficgler, H., Giresi, P.G., Goldy, J., Hawrylycz, M., Haydock, A., Humbert, R., James, K.D., Johnson, B.E., Johnson, E.M., Frum, T.T., Rosenzweig, E.R., Karnani, N., Lee, K., Lefebvre, G.C., Navas, P.A., Neri, F., Parker, S.C., Sabo, P.J., Sandstrom, R., Shafer, A., Vetrie, D., Weaver, M., Wilcox, S., Yu, M., Collins, F.S., Dekker, J., Lieb, J.D., Tullius, T.D., Crawford, G.E., Sunyaev, S., Noble, W.S., Dunham, I., Denoeud, F., Reymond, A., Kapranov, P., Rozowsky, J., Zheng, D., Castelo, R., Frankish, A., Harrow, J., Ghosh, S., Sandelin, A., Hofacker, I.L., Baertsch, R., Keefe, D., Dike, S., Cheng, J., Hirsch, H.A., Sekinger, E.A., Lagarde, J., Abril, J.F., Shahab, A., Flamm, C., Fried, C., Hackermüller, J., Hertel, J., Lindemeyer, M., Missal, K., Tanzer, A., Washietl, S., Korbel, J., Emanuelsson, O., Pedersen, J.S., Holroyd, N., Taylor, R., Swarbreck, D., Matthews, N., Dickson, M.C., Thomas, D.J., Weirauch, M.T., Gilbert, J., Drenkow, J., Bell, I., Zhao, X., Srinivasan, K.G., Sung, W.K., Ooi, H.S., Chiu, K.P., Foissac, S., Alioto, T., Brent, M., Pachter, L., Tress, M.L., Valencia, A., Choo, S.W., Choo, C.Y., Ucla, C., Manzano, C., Wyss, C., Cheung, E., Clark, T.G., Brown, J.B., Ganesh, M., Patel, S., Tammana, H., Chrast, J., Henrichsen, C.N., Kai, C., Kawai, J., Nagalakshmi, U., Wu, J., Lian, Z., Lian, J., Newburger, P., Zhang, X., Bickel, P., Mattick, J.S., Carninci, P., Hayashizaki, Y., Weissman, S., Hubbard, T., Myers, R.M., Rogers, J., Stadler, P.F., Lowe, T.M., Wei, C.L., Ruan, Y., Struhl, K., Gerstein, M., Antonarakis, S.E., Fu, Y., Green, E.D., Karaöz, U., Siepel, A., Taylor, J., Liefer, L.A., Wetterstrand, K.A., Good, P.J., Feingold, E.A., Guyer, M.S., Cooper, G.M., Asimenos, G., Dewey, C.N., Hou, M., Nikolaev, S., Montoya-Burgos, J.I., Löytynoja, A., Whelan, S., Pardi, F., Massingham, T., Huang, H., Zhang, N.R., Holmes, I., Mullikin, J.C., Ureta-Vidal, A., Paten, B., Seringhaus, M., Church, D., Rosenbloom, K., Kent, W.J., Stone, E.A., Batzoglou, S., Goldman, N., Hardison, R.C., Haussler, D., Miller, W., Sidow, A., Trinklein, N.D., Zhang, Z.D., Barrera, L., Stuart, R., King, D.C., Ameur, A., Enroth, S., Bieda, M.C., Kim, J., Bhinge, A.A., Jiang, N., Liu, J., Yao, F., Vega, V.B., Lee, C.W., Ng, P., Shahab, A., Yang, A., Moqtaderi, Z., Zhu, Z., Xu, X., Squazzo, S., Oberley, M.J., Inman, D., Singer, M.A., Richmond, T.A., Munn, K.J., Rada-Iglesias, A., Wallerman, O., Komorowski, J., Fowler, J.C., Couttet, P., Bruce, A.W., Dovey, O.M., Ellis, P.D., Langford, C.F., Nix, D.A., Euskirchen, G., Hartman, S., Urban, A.E., Kraus, P., Van Calcar, S., Heintzman, N., Kim, T.H., Wang, K., Qu, C., Hon, G., Luna, R., Glass, C.K., Rosenfeld, M.G., Aldred, S.F., Cooper, S.J., Halees, A., Lin, J.M., Shulha, H.P., Zhang, X., Xu, M., Haidar, J.N., Yu, Y., Ruan, Y., Iyer, V.R., Green, R.D., Wadelius, C., Farnham, P.J., Ren, B., Harte, R.A., Hinrichs, A.S., Trumbower, H., Clawson, H., Hillman-Jackson, J., Zweig, A.S., Smith, K., Thakkapallayil, A., Barber, G., Kuhn, R.M., Karolchik, D., Armengol, L., Bird, C.P., de Bakker, P.I., Kern, A.D., Lopez-Bigas, N., Martin, J.D., Stranger, B.E., Woodroffe, A., Davydov, E., Dimas, A., Eyras, E., Hallgrímsdóttir, I.B., Huppert, J., Zody, M.C., Abecasis, G.R., Estivill, X., Bouffard, G.G., Guan, X., Hansen, N.F., Idol, J.R., Maduro, V.V., Maskeri, B., McDowell, J.C., Park, M., Thomas, P.J., Young, A.C., Blakesley, R.W., Muzny, D.M., Sodergren, E., Wheeler, D.A., Worley, K.C., Jiang, H., Weinstock, G.M., Gibbs, R.A., Graves, T., Fulton, R., Mardis, E.R., Wilson, R.K., Clamp, M., Cuff, J., Gnerre, S., Jaffe, D.B., Chang, J.L., Lindblad-Toh, K., Lander, E.S., Koriabine, M., Nefedov, M., Osoegawa, K., Yoshinaga, Y., Zhu, B., de Jong, P.J. ENCODE Project Consortium; (2007). NISC Comparative Sequencing Program; (2007). Baylor College of Medicine Human Genome Sequencing Center; (2007). Washington University Genome Sequencing Center; (2007). Broad Institute; (2007). Children's Hospital Oakland Research Institute. (2007). Identification and analysis of functional elements in 1% of the human genome by the ENCODE pilot project. *Nature, 447*(7146), 799-816.
[http://dx.doi.org/10.1038/nature05874] [PMID: 17571346]

Ng, P., Wei, C.L., Sung, W.K., Chiu, K.P., Lipovich, L., Ang, C.C., Gupta, S., Shahab, A., Ridwan, A., Wong, C.H., Liu, E.T., Ruan, Y. (2005). Gene identification signature (GIS) analysis for transcriptome characterization and genome annotation. *Nat. Methods, 2*(2), 105-111. [http://dx.doi.org/10.1038/nmeth733] [PMID: 15782207]

CHAPTER 11

Single Cell Sequencing (SCS) and Single Cell Transcriptome (SCT) Sequencing

Abstract: Sometimes genomic and transcriptomic information of single cells, instead of those produced from cell populations, are desired. Obtaining such information relies on single cell sequencing (SCS) and single cell transcriptome (SCT) sequencing.

Although SCT sequencing is in fact part of SCS, they are readily distinguishable not only in research objective, but also in experimental procedure and bioinformatic approach. We will first review the history and achievements that have been made in these fields, and then discuss an experimental procedure of SCT sequencing to gain more insight into the subject.

Keywords: SCS, Single cell Sequencing, Single cell transcriptome sequencing, SCT, Transcriptome.

INTRODUCTION

Single cell sequencing (SCS) refers to the sequencing and analysis of the genomic or transcriptomic sequences of single prokaryotic or eukaryotic cells. It possesses unprecedented potential in resolving genetic substructures and the variations in genomic and transcriptomic profiles at both single cell and molecular levels. During the past few years, studies on SCS have shown promising results, especially in cancer research and transcriptome analysis. Foreseeing its great potential applications in biological studies, SCS was chosen as Method of the Year in 2013 by *Nature Methods*.

Current SCS approaches rely on both PCR-based DNA or cDNA amplification and next-generation sequencing. NGS is required because it is the only sequencing approah allowing us to obtain an in-depth coverage. Conventional NGS technologies, however, require large amounts of genetic material (a few nanograms or micrograms per library) as the input for sequencing. Such quantities are many orders of magnitude higher than a single cell can provide, making PCR amplification also an essential element for single cell sequencing.

So far, SCS applications mainly focus on single cell genome sequencing and single cell transcriptome sequencing. The former was highlighted by the pilot studies on cancer evolution by Navin and Hou *et al.* (Hou *et al.*, 2012; Navin *et al.*, 2011), while the latter by the works on development and cancer transcriptome by Tang and Ramskold *et al.*

Kuo Ping Chiu

(Ramskold *et al.*, 2012; Tang *et al.*, 2009). In the paper published in 2011, Navin and colleagues reported their work of using single nucleus sequencing (SNS) to study copy number variation in breast cancer (Navin *et al.*, 2011). By using single cells from the same cancer origin, they were able to determine the genetic lineages, and thus the evolutionary substructures, in a cancer. Their results suggested that tumors grow by punctuated clonal expansion, instead of gradual tumor progression. The next year, Hou and colleagues published the multiple displacement amplification (MDA) method for single cell genomic DNA amplification and its application in the study of the genes involved in essential thrombocythemia (ET) evolution (Hou *et al.*, 2012). Results suggested that ET patient carries a distinct set of mutations and a monoclonal origin of ET cancer cells.

For single cell transcriptome analysis, certain methods to improve single cell cDNA amplification were already reported prior to the advent of NGS technologies. These include the work published by Eberwine *et al.*, in 1992 and that published by Kurimoto *et al.*, in 2006 (Eberwine *et al.*, 1992; Kurimoto *et al.*, 2006). The NGS-based single cell transcriptome sequencing was first published in 2009 by Tang *et al.*, (Tang *et al.*, 2009). In a work published by Ramskold *et al.*, in 2012, they reported an elegant method, called Smart-Seq, for reverse transcription and cDNA amplification for SCT analysis (Ramskold ., 2012). The method has been commercialized by Clontech in making "the SMARTer Ultra Low RNA Kit". For technical control, they used defined quantities of total RNA, which could be correlated to defined numbers of cells. In fact, to compensate insufficient qantity of input material for next-generation sequencing, we have been rountinely using defined amounts of total RNA or mRNA for transcriptome analyses since years ago. Our results demonstrated the feasibility of this modification and also indicated that, as expected, mRNA works better than total RNA.

Most people have a concern about the technical variations which may impose bias to SCT analysis. Certainly, many factors may introduce bias into single cell transcriptome analysis. SCT bias may result from variations in personal technical skills, and may confuse the real transcriptional variation among individual cells (Ramskold *et al.*, 2012). To minimize the influence of variability in personal technical skill, increasing the number of SCTs and using internal controls such as house keeping genes and previously studied expression patterns of certain genes is recommended.

Experimental procedure for SCT sequencing

To help readers further understand how SCS is conducted, here I would like to present and discuss the experimental procedure that we are using in the lab. Since SCS per se is a broad subject, it would be better off for us to limit the scope by focusing on SCT analysis. For SCS applications in the study of genetic sequence variations, cancer gene identification and cancer evolution, please consult the works published by Navin and

How *et al.*, For more useful information about SCT experiments please consult the original procedure published by Ramskold and colleagues (Ramskold *et al.*, 2012). We have compared Ramskold's method with others and found that this method is the most reproducible. The experimental procedure outlined below mainly follows this method but with certain modifications.

Basically, the procedure can be divided into two parts: 1) generation and amplification of cDNA and 2) sequencing library construction. We will go through the general ideas. Please consult the original procedure for more detailed information.

Generation and amplification of cDNA

Cells are trypsinized (if needed), washed, and kept in a buffer (*e.g.* PBS) before single cell isolation. Single cell isolation can be done by mouth pipetting or by other methods of micromanipulation. We prefer mouth pipetting because it is very mild and does not cause cell damage. Each single cell is first kept in ≤ 1 uL of 1X PBS and then lysed by adding 4 uL of hypotonic lysis buffer which contains RNase inhibitors together with other ingredients (see the original protocol). The first strand cDNA synthesis is primed by CDS primer (5′–AAGCAGTGGTATCAACGCAGAGTACT(30)VN–3′, where 'V' stands for non-T). The degenerated base (V) is so designed to enhance binding of the oligo to the beginning of the polyA stretch. Without it, the oligo may "slip" within the polyA region. The MMLV reverse transcriptase (RT) possesses a terminal transferase activity and would automatically add a few extra C (polyC tail) to the 3′ end of the first strand cDNA. The reaction solution also contains SMARTer II A oligo (5′-AAGCAGTGGTATCAA-CGCAGAGTACATrGrGrG-3′, where 'r' stands for ribonucleotide base) which is able to anneal to the polyC tail in the first strand cDNA, the SMARTer II A oligo thus creates an extended template, allowing the MMTV RT to continue the first strand cDNA synthesis throughout the second template.Since the sequence in the SMARTer II A oligo is known, PCR primers are designed and used not only to make the second strand of cDNA, but also for the amplification of cDNA.

For "multiple-cell" transcriptome analyses, MCF-7 cDNA was pre-amplified for 12 cycles when using 1 ng of total RNA (~100 MCF-7 cells), or 15 cycles when using 100 pg of total RNA (~10 MCF-7 cells), or 18 cycles when using 10 pg of total RNA (~1 MCF-7 cell). The cDNA samples prepared from MCF-10A or MCF-7 single cells were amplified for 23 cycles and 20 cycles, respectively, because MCF-10A transcribes much less RNA than MCF-7. The PCR-amplified cDNA should appear as a distinct peak of ~500–5,000 bp under Bioanalyzer (Fig. 1).

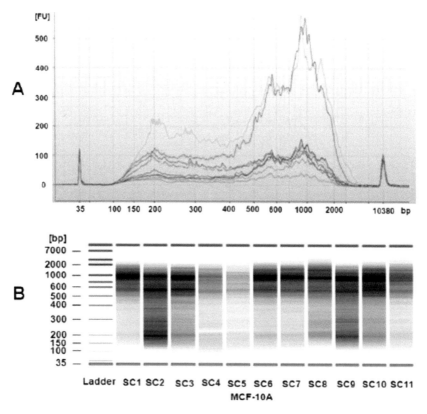

Fig. (1). Eleven MCF-10A single cells were isolated, lyzed and subjected to cDNA synthesis and amplification using SMARTer Ultra Low RNA Kit. As shown in the figure by Bioanalyzer, the size of cDNA falls within a well-defined range.

Construction and sequencing of Smart-Seq sequencing libraries

The amplified cDNA samples are then subjected to sequencing library construction for Illumina barcoded paired-end (PE) sequencing. Before being ligated to the sequencing adaptors, cDNA is fragmented by sonicator (*e.g.* Covaris S2). cDNA fragments of a few hundred bp in size (*e.g.* ~150 - 400bp) are selected by agarose gel excision or by beads. Amplified cDNA (~5 ng) is used to construct an Illumina sequencing library. Fragmentation leaves uneven termini in the cDNA fragments, so that the cDNA sample has to be end-repaired. Then, an 'A' nucleotide is added to the 3' end of the cDNA fragments, making the sample ready for single-base sticky-end ligation with Illumina pair-end adaptors with barcodes/indexes (*e.g.* 12 barcodes were used for our MCF-10A libraries and 6 barcodes used for our MCF-7 libraries, one for each). After ligation, SPRI XP Beads were used to purify the library (by removing residual adaptors). The sample is amplified by PCR for 12–18 cycles and purified by SPRI XP Beads again to remove residual primers. DNA in the library is quantified using Bioanalyzer (Agilent) or Fragment Analyzer™ (Advanced Analytical Technologies, Inc. (AATI)). The latter is recommended because it is more cost-effective and efficient. An example is shown in Fig. (2).

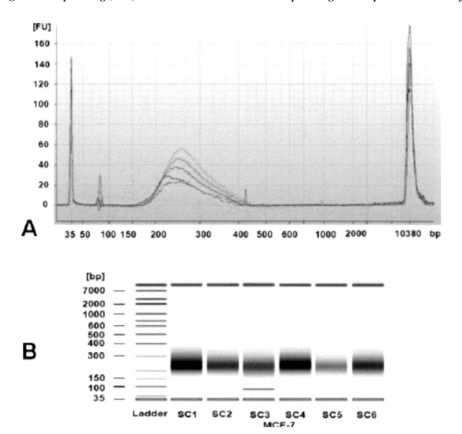

Fig. (2). cDNA samples of six MCF-7 single cells were sonicated, end repaired/A-tailed, P5/P7 adaptor-ligated, size selected, and PCR amplified for 18 cycles prior to Bioanalyzer display. As shown in the figure, the majority of DNA constructs have sizes ranging between 150-400 bp.

Fig. (3). MiSeq run summary of 12 single cell libraries.

Before loading to the sequencer, the library needs to be quantified accurately by qPCR machine (*e.g.* Roche LightCycler "480 II"). The concentration is adjusted to 12 pM and then pooled together for sequencing by Illumina HiSeq 2000, GAIIx or MiSeq sequencer. A run summary by MiSeq is shown in Fig. (3).

Modifications made on single cell transcriptome (SCT) libraries

Protocols may need modification for a number of reasons. First, a protocol may not have been optimized befor it was published. This frequently occurs with protocols published in a hurry by a company trying to catch up with market demand. Besides having prematurely rushed to meet market pressures, there can be other good reasons for us to change or improve a protocol. Here, I will use single cell transcriptome (SCT) analysis as an example and go through some potential changes we can make on a wetlab protocol.

A. Using different types of material as the input

Instead of using single cells, one can use either total RNA or mRNA as the input material for the SCT protocol. mRNA molecules are included in the total RNA population and are the sole molecules from the single cell that we want to convert into double-stranded cDNAs and amplify by PCR for transcriptome analysis. Thus, it is reasonable to use either total RNA or mRNA to replace single cells.

The majority of RNA molecules are ribosomal RNA (rRNA) and transfer RNA (tRNA), while mRNA species constitute only ~2% of the total RNA. The presence of rRNA and tRNA may reduce the efficiency of a number of reactions, including oligo-dT priming, reverse transcription, and PCR amplification, *etc*. As such, it is strongly recommended to use mRNA as the input material, if the amount of total RNA is sufficient for mRNA isolation. In fact, it has been empirically demonstrated that mRNA works better than total RNA.

How, then, do we correlate the results produced from mRNA or total RNA with cell number? Based on a rough estimation using MCF-7 breast cancer cells, each MCF-7 cell expresses about 10 pg of total RNA. Accordingly, 50 ng of total RNA is equivalent to about 5000 cells, and 50 ng of mRNA is equivalent to 250,000 (= 0.25 million) single cells. You can calculate and use a certain amount of total RNA or mRNA based on the number of cells you want to use in your study. Since the total amount of RNA expressed from a single cell varies across different cell types, you are strongly recommended to empirically fine tune this value based on the cell type that you are using.

B. Using different primers to prime cDNA synthesis

Tang *et al.*, (2009) used oligo-dT primers without a VN tail (where 'V' represents a non-T nucleotide (either A, G, or C) and 'N' represents either A, T, G, or C, for cDNA synthesis). Alternatively, one could choose to use oligo-dT primers containing a VN tail to prime cDNA synthesis. Since the lengths of the polyA regions among mRNA molecules vary significantly and some can be much longer than the primer, priming by non-VN primers may shift along the polyA region within each mRNA. This problem can result in a DNA population with high length variation and cause misinterpretation of the transcriptome profile. The presence of VN at the 3' end of an oligo-dT primer should be able to increase the specificity of cDNA synthesis.

C. Choosing shotgun fragment sequencing or Paired-End Ditag sequencing

Completion of the 2nd PCR amplification marks the junction where the SCT protocol and other protocols meet, because the retrieval of mRNA molecular information from a single cell is completed and preparation of a sequencing library can be initiated. Here, one can determine to take the shotgun fragment sequencing approach (followed by RNA-Seq analysis), or the Pair-End Ditag (PED) sequencing approach (followed by PED analysis).

Sequence processing and analysis

Sequencing processing and analysis are the same as in the case of whole transcriptome sequencing/analysis.

REFERENCES

Eberwine, J., Yeh, H., Miyashiro, K., Cao, Y., Nair, S., Finnell, R., Zettel, M., Coleman, P. (1992). Analysis of gene expression in single live neurons. *Proc. Natl. Acad. Sci. USA, 89*(7), 3010-3014.
[http://dx.doi.org/10.1073/pnas.89.7.3010] [PMID: 1557406]

Hou, Y., Song, L., Zhu, P., Zhang, B., Tao, Y., Xu, X., Li, F., Wu, K., Liang, J., Shao, D., Wu, H., Ye, X., Ye, C., Wu, R., Jian, M., Chen, Y., Xie, W., Zhang, R., Chen, L., Liu, X., Yao, X., Zheng, H., Yu, C., Li, Q., Gong, Z., Mao, M., Yang, X., Yang, L., Li, J., Wang, W., Lu, Z., Gu, N., Laurie, G., Bolund, L., Kristiansen, K., Wang, J., Yang, H., Li, Y., Zhang, X., Wang, J. (2012). Single-cell exome sequencing and monoclonal evolution of a JAK2-negative myeloproliferative neoplasm. *Cell, 148*(5), 873-885.
[http://dx.doi.org/10.1016/j.cell.2012.02.028] [PMID: 22385957]

Kurimoto, K., Yabuta, Y., Ohinata, Y., Ono, Y., Uno, K.D., Yamada, R.G., Ueda, H.R., Saitou, M. (2006). An improved single-cell cDNA amplification method for efficient high-density oligonucleotide microarray analysis. *Nucleic Acids Res., 34*(5), e42.
[http://dx.doi.org/10.1093/nar/gkl050] [PMID: 16547197]

Navin, N., Kendall, J., Troge, J., Andrews, P., Rodgers, L., McIndoo, J., Cook, K., Stepansky, A., Levy, D., Esposito, D., Muthuswamy, L., Krasnitz, A., McCombie, W.R., Hicks, J., Wigler, M. (2011). Tumour evolution inferred by single-cell sequencing. *Nature, 472*(7341), 90-94.
[http://dx.doi.org/10.1038/nature09807] [PMID: 21399628]

Ramsköld, D., Luo, S., Wang, Y.C., Li, R., Deng, Q., Faridani, O.R., Daniels, G.A., Khrebtukova, I., Loring, J.F., Laurent, L.C., Schroth, G.P., Sandberg, R. (2012). Full-length mRNA-Seq from single-cell levels of RNA and individual circulating tumor cells. *Nat. Biotechnol., 30*(8), 777-782.
[http://dx.doi.org/10.1038/nbt.2282] [PMID: 22820318]

Tang, F., Barbacioru, C., Wang, Y., Nordman, E., Lee, C., Xu, N., Wang, X., Bodeau, J., Tuch, B.B., Siddiqui, A., Lao, K., Surani, M.A. (2009). mRNA-Seq whole-transcriptome analysis of a single cell. *Nat. Methods, 6*(5), 377-382.
[http://dx.doi.org/10.1038/nmeth.1315] [PMID: 19349980]

CHAPTER 12

ChIP-TFBS Analysis

Abstract: Eukaryotic gene expression is tightly controlled by a cascade of regulatory mechanisms. At the sequence level, gene expression is regulated by *cis*-acting DNA motifs that are able to recruit trans-acting transcription factors (TFs) for positive or negative regulation of local gene expression. The genome-wide mapping of transcription factor binding sites (TFBS) becomes a crucial strategy for the study of gene expression regulation. Here in this chapter we will discuss the preparation of ChIP-TFBS sequencing libraries and the analysis of ChIP-TFBS sequence data.

Keywords: ChIP, ChIP-TFBS, Chromatin immunoprecipitation, Motif, TF, Transcription factor.

Definition of Terminologies

Transcription Factor (TF)

A transcription factor is a protein that binds to specific DNA motifs in the genome and works together with other proteins (including co-factors, helicase and RNA polymerase) to enhance, or block, the transcription (expression) of genes.It is a well-studied mechanism for the regulation of gene expression.

TFBSs: *Transcription factor binding sites*

INTRODUCTION

ChIP-TFBS (Chromatin IP-mediated transcription factor binding site) analysis is a common practice in the study of transcriptional regulation (Park, 2009; Pepke *et al.*, 2009). (Lin *et al.*, 2007; Loh *et al.*, 2006; Zeller *et al.*, 2006). Frequently, transcriptional regulation of gene expression is initiated by ligand binding to target receptors on a cell membrane followed by a cascade of molecular signaling to induce transcriptional activation of a group of genes and a simultaneous repression of another group of genes. In other words, ligands, such as hormones or cytokines, bind to their target receptors on a cell membrane to initiate a cascade of signal transduction, leading to the activation of one or more specific TF, which enter(s) the nucleus and bind(s) to its/their accessible motifs in the genome. A TF, by recruiting its co-factors and transcriptional machinery, in turn activates the expression of one group of target genes and simultaneously represses another group of genes. During this process, TF binding plays a key role in determining the specificity for both transcriptional activation and repression.

Kuo Ping Chiu

For TFBS analysis, chromosomes need to be fragmented into smaller pieces. However, prior to sequencing library construction, regional contacts between DNA and proteins need to be maintained. Moreover, to date ways to probe transcription factor binding sites (which are DNA or RNA sequences dispersed across the genome) without using antibody (Ab) to "fish" the DNA-protein complexes of interest have not been developed. To address these various complications, we need to fix cells (either in culture, tissue section, or other kind of cell sample), isolate the chromatin (containing both DNA and associated proteins), fragment the chromatin, enrich the chromatin fragments of interest, then remove proteins and construct a sequencing library on the DNA fragments that are hypothesized to be previously bound by the transcription factor under investigation. This process is shared by ChIP-EM (epigenetic modification) which will be introduced in the nextchapter.

Experimental Procedure

First of all, trypsinized cultured cells or tissue samples must be fixed by a cross-linking fixative (Fig. 1). Precipitative fixatives are not suitable for this purpose. Normally, ~0.5-2% paraformaldehyde, or formalin, is used. This fixation step is intended to preserve the original state of the chromatin. After fixation, cells can be collected into a test tube for chromatin isolation. Once the chromatin is collected, it is fragmented into smaller pieces, usually a few hundred to a few thousand base pairs in size. For this purpose, sonication, instead of enzymatic digestion, is recommended, as the former generates chromatin fragments by random breakage caused by mechanical force, while the latter produces RE site-defined fragments. To minimize bias and artifacts, chromatin fragments resulting from random breakage are preferred.

After sonication, the desired chromatin fragments are enriched by a specific antibody which recognizes, and binds to, the TF of interest. (Remember, only fixed chromatin fragments can be enriched by an antibody!). ChIP-TFBS analysis relies on antibodies against specific epitopes in the transcription factor. This step enriches TF-bound ChIP fragments. All proteins are then erased (removed) from the enriched DNA fragments, which are subsequently end-repaired, tailed with an 'A', ligated to sequencing adaptors, denatured, annealed to anchored oligos for *in situ* PCR amplification, and then subjected to sequencing. A ChIP-TFBS library can be sequenced as a (shotgun) fragment library, a PE (paired-end) library, or a PED (paired-end ditag) library, depending on the investigator's preference. However, to avoid confusion resulting from low coverage, a PED library is strongly recommended. Sequences produced by the sequencer are expected to harbor the genomic locations bound by the TF under investigation.

The key to the success of ChIP-TFBS analysis relies on using the appropriate fixatives and antibodies at the appropriate concentrations and under the appropriate conditions. The cross-linking fixative preserves DNA/RNA and bound proteins in their original locations, but still allows the bound proteins to be removed/degraded in a later step. It is not difficult to determine which fixative to use. There are two types of fixatives (either cross-linking fixatives or precipitative fixatives) and the cross-linking fixative paraformaldehyde has long been known to best serve this purpose. On the other hand, it can be difficult to define the optimal fixation conditions because the efficiency of fixation is affected not only by sample type (*e.g.*, tissue culture, tissue section, *etc.*) and material sources (*e.g.*, microorganism, fish, human, *etc.*), but also by the thickness and conditions of preservation (*e.g.*, fresh tissue section, liquid nitrogen-preserved tissue section, paraffin-embedded tissue sections, *etc.*). As such, optimization of fixation conditions may be required for each experiment. Similarly, the antibody chosen plays a key role in determining the specificity of binding. Since the specificity is influenced by a number of factors (including the Ab itself, its concentration, salt concentration, temperature and the stringency of wash conditions, *etc.*), optimization of hybridization and wash conditions are normally required for each antibody.

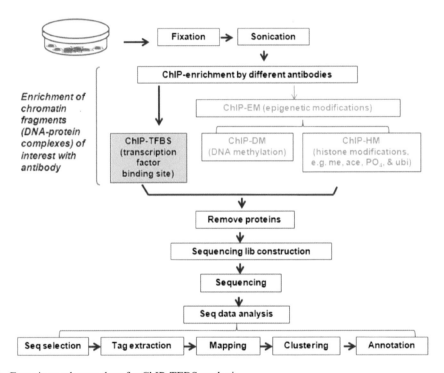

Fig. (1). Experimental procedure for ChIP-TFBS analysis.

Sequence Data Analysis

The initial sequence processing is basically the same for all types of sequence data. That is, the raw sequences need to be cleaned up (including decontamination; trimming of vector sequences, if any; removal of low-quality regions/sequences, *etc.*) and sometimes re-organized. The quality reads are then mapped against the reference genome to identify their genomic origins (chromosomal locations). Aligned reads are grouped into clusters based on their chromosomal locations. Each cluster is expected to be a binding site of that transcription factor. Generally speaking, a few hundred to a few thousand binding sites can be found for each transcription factor, depending on the transcription factor under study and the experimental and analytical stringency. Normally, singletons (*i.e.*, locations each mapped by a single read) are considered as non-specific and must be excluded from analysis.

OBSERVATIONS AND DISCUSSION

Transcription Factor Binding Sites Can Be Well Represented Using the Paired-end Ditag Approach

As mentioned above, a ChIP-TFBS library can be sequenced as a fragment library, a PE library, or a PED library. However, a PED library is preferred because the boundaries can be clearly defined (Fig. **2**).

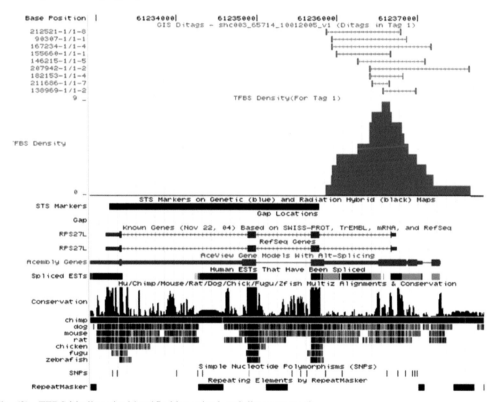

Fig. (2). TFBS binding site identified by paired-end ditag approach.

TFBS Clusters are Well Correlated With Gene-Rich Regions

By displaying the binding sites of multiple transcription factors in parallel, it is easy to see that clusters of TFBSs correlate very well with gene-rich regions in the genome (Figs. **3** and **4**). In fact, this is expected, because transcription factors regulate the transcription of genes and most of their binding sites should be in close proximity to genes, although some may reside in intergenic regions.

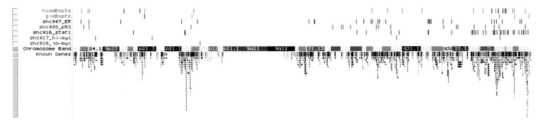

Fig. (3). Binding of some key transcription factors in chromosome 9.

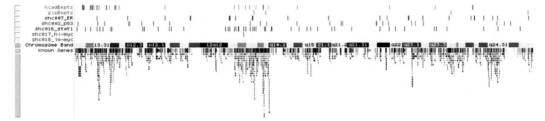

Fig. (4). Binding of some key transcription factors in chromosome 12.

Consensus Motif Sequence can be Extracted from Sequences Bound by a Transcription Factor Using Computer Software such as GLAM or MIME

From transcription factor-bound sequences, one can extract a consensus motif sequence and display it in a sequence logo (Fig. **5**), which is a graphic representation of the sequence conservation for every nucleotide in the motif. It is also possible to compare the consensus motif sequence with a previously discovered sequence or to conduct cross-library comparisons.

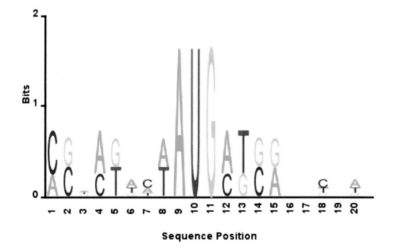

Fig. (5). A sequence logo showing the most conserved bases around the initiation codon from all human mRNAs.

As shown in Table **1**, TFBSs can be categorized into intergenic, intragenic (including exonic and intronic), 5'-distal (*e.g.*, 100 – 10 Kb from the transcription start site (TSS)), 5'-proximal (*e.g.*, ≤ 10 Kb from TSS), 3'-distal (*e.g.*, 100 - 10 Kb from the polyadenylation site (PAS)), 3'-proximal (*e.g.*, ≤ 10 Kb from PAS), promoter-associated, or enhancer-associated, *etc*. Interestingly, many of the TFBSs studied so far reside in the first intron, suggesting that the first intron may play an important role in gene expresson regulation.

Table 1. Mapping of Oct4 and Nanog binding sites in the mouse ES genome and p53 and cMyc in human cancer genomes.

TF	Cutoff	# Clusters	5' distal (100-10 Kb)	5' proximal (<10 Kb)	intragenic intron1	intron1>1	exonic	3' proximal (<10 Kb)	3' distal (100-10 Kb)	Desert region (>100 Kb)
Oct4 Mouse ES cells	≥ 4 ditags per cluster (4+)	Total **1,083** analyzed	13.1%	18.6%	41.2%		2.3%	7.5%	9.8%	7.5%
			31.7%		43.5%			17.3%		
Nanog Mouse ES cells	≥ 4 ditags per cluster (4+)	Total **3,006** analyzed	21.31%	12.8%	31.4%		1.9%	7.4%	17.9%	7.3%
			34.1%		333%			23.3%		
p53 Human HCT116	2+/3+	Total **542** analyzed	156 (28.8%)		46	152	0	120 (22.1%)		68 (12.5%)
					198 (36.5%)					
cMyc Human B lymphoma	≥3 ditags per cluster (3+)	Total **610** analyzed	94	130	86	148	0	99		53 (8.7%)
			224 (36.7%)		234 (38.4%)			77 (16.2%)		

There can be Significant Overlap Between the Binding Sites of Transcription Factors with Closely Related Functions

Transcription factors with similar or closely related functions may bind to the same, or similar, genomic locations. Here, we use Nanog and Oct4 as an example to demonstrate this possibility (Fig. **6**).

Fig. (6). Colocalization of Oct4 and Nanog binding sites in mouse ES cells.

REFERENCES

Lin, C.Y., Vega, V.B., Thomsen, J.S., Zhang, T., Kong, S.L., Xie, M., Chiu, K.P., Lipovich, L., Barnett, D.H., Stossi, F., Yeo, A., George, J., Kuznetsov, V.A., Lee, Y.K., Charn, T.H., Palanisamy, N., Miller, L.D., Cheung, E., Katzenellenbogen, B.S., Ruan, Y., Bourque, G., Wei, C.L., Liu, E.T. (2007). Whole-genome cartography of estrogen receptor alpha binding sites. *PLoS Genet., 3*(6), e87.
[http://dx.doi.org/10.1371/journal.pgen.0030087] [PMID: 17542648]

Loh, Y.H., Wu, Q., Chew, J.L., Vega, V.B., Zhang, W., Chen, X., Bourque, G., George, J., Leong, B., Liu, J., Wong, K.Y., Sung, K.W., Lee, C.W., Zhao, X.D., Chiu, K.P., Lipovich, L., Kuznetsov, V.A., Robson, P., Stanton, L.W., Wei, C.L., Ruan, Y., Lim, B., Ng, H.H. (2006). The Oct4 and Nanog transcription network regulates pluripotency in mouse embryonic stem cells. *Nat. Genet., 38*(4), 431-440.
[http://dx.doi.org/10.1038/ng1760] [PMID: 16518401]

Park, P.J. (2009). ChIP-seq: advantages and challenges of a maturing technology. *Nat. Rev. Genet., 10*(10), 669-680.
[http://dx.doi.org/10.1038/nrg2641] [PMID: 19736561]

Pepke, S., Wold, B., Mortazavi, A. (2009). Computation for ChIP-seq and RNA-seq studies. *Nat. Methods, 6*(11) (Suppl.), S22-S32.
[http://dx.doi.org/10.1038/nmeth.1371] [PMID: 19844228]

Zeller, K.I., Zhao, X., Lee, C.W., Chiu, K.P., Yao, F., Yustein, J.T., Ooi, H.S., Orlov, Y.L., Shahab, A., Yong, H.C., Fu, Y., Weng, Z., Kuznetsov, V.A., Sung, W.K., Ruan, Y., Dang, C.V., Wei, C.L. (2006). Global mapping of c-Myc binding sites and target gene networks in human B cells. *Proc. Natl. Acad. Sci. USA, 103*(47), 17834-17839.
[http://dx.doi.org/10.1073/pnas.0604129103] [PMID: 17093053]

CHAPTER 13

ChIP-EM Libraries

Abstract: Epigenetic modifications (EMs) refer to the external modifications on DNA that do not alter coding specificity. EMs include DNA methylations (DMs) and histone modifications (HMs). This chapter will focus on HMs. We will discuss how ChIP-EM libraries can be made and what can be expected from the sequence data analysis. There are many laboratories working in this field and many reports have been published. Readers are recommended to consult the previous reports for further understanding of this subject.

Keywords: Chromatin immunoprecipitation, ChIP, EM, Epigenetic modifications.

Definition of Terminologies
Epigenetic Modifications
Epigenetic modifications (EM) refer to the mitotically and/or meiotically heritable, but biochemically reversible, on-chromatin modifications, which confer phenotypic plasticity by coordinating the expression of multiple genes in a 3-dimensionally dispersed but functionally correlated manner without entailing any change in DNA sequence. These modifications include 1) the use of different histone variants, 2) the methylation, acetylation, phosphorylation, ubiquitination, or sumoylation of nucleosomal histone proteins, and 3) DNA methylations. This terminology was first described by Conrad Waddington in his paper entitled "The epigenotype" published in 1942 (Waddington, 2012).

INTRODUCTION

Chromatin structures are molecular complexes made primarily of both DNA and proteins. Histones are the most prevalent protein species in chromatin. By forming positively charged histone cores, each of which is made up of 2 copies of (H2AH2BH3H4) and left-handedly wrapped by ~147 bp of negatively charged DNA, histone proteins play a key role in DNA packaging. As noted in Chapter 1, there are three types of chromatin - heterochromatin, euchromatin, and facultative euchromatin - which occupy distinctive segments and form scattered euchromatin (loosely packaged) or heterochromatin (densely packaged) islands in the nuclear genome. Certain amino acid residues in the N-terminal protrusions, or so-called "tails", of histones are posttranslationally modified by a few types of small moieties. These modifications, including acetylation, methylation, phosphorylation, ubiquitination, sumoylation, ADP ribosylation, deamination, and proline isomerization, affect the degree of DNA packaging and thus influence DNA binding by various proteins (including TFs, cofactors, RNA polymerase, helicase, and topoisomerase), which in turn influences the expression of genes in the modification regions. In fact, functional characterization has also implicated histone modifications in multiple biological processes, including DNA replication, DNA repair, apoptosis, embryogenesis, cell cycle regulation, and embryonic and neuronal development (Arnaudo and Garcia, 2013; Graff *et al.*, 2011; Hirabayashi and Gotoh, 2010; Kouzarides, 2007). There are a total of about 60 residues in each histone core which can be modified by at least one moiety per site. This raises the question of how such complex modifications in nucleosomal histones regulate gene expression and participate in so many biological functions.

THE LAW OF UNCERTAINTY AT THE EPIGENETIC MODIFICATION LEVEL

In physics, Heisenberg's uncertainty principle follows the fomula $SD_x \times SD_y \geq h/2$, where SD_x stands for the standard deviation of position, SD_y stands for the standard deviation of momentum, and h stands for Plank's constant. Uncertainty exists in every object under study due to the influence introduced by the interaction between objects. When extrapolated from the quantum level to the molecular level, and further to the cellular level or higher, the law of uncertainty in physics introduces uncertainty to the biological system as all biological phenomena obey chemical and physical laws. Since we cannot precisely define the position of an electron, we cannot precisely define the shape or position of a protein. In the strict sense, we more or less adhere to the concept of probability when we describe a biological information.

In Biology, the Law of Uncertainty is Shown in Genetic Mutations (*e.g.*, SNVs), as Well as at the Level of Epigenetic Modifications

Epigenetic modifications (as well as TFBSs) are very dynamic. Moreover, there are antagonistic and synergistic interactions between epigenetic modifications and, with so many modifiable amino acid residues (~60 per nucleosome) and some residues that may have multiple types of modifications, the histone modifications alone are both spacially and temporally overcrowded, further amplifying the antagonistic and synergistic effects among EMs. Antibodies and mass spectrometry are the most commonly used methods for studying EMs. Like all other methods, both of these methods have their intrinsic limitations. For example, antibodies have their limitations (uncertainties) in specificity and sensitivity, while mass spectrometry has its limitations (uncertainties) in fragmentation and sensitivity. Both limit the accuracy of EM investigations.

DNA methylations occur most frequently in the cytosines of CpG islands, some of which are found in promoter regions of genes and some in the intragenic or intergenic regions.

CpG islands are typically about 300 – 3,000 bp in size. About 40% of mammalian genes have CpG islands in the promoter, some of which are methylated. DNA methylation in CpG islands has frequently been found to result in transcriptional repression, although there are some obvious exceptions. Cross-talk between DNA methylation and histone modification has also been observed (Cedar and Bergman, 2009; Fischle *et al.*, 2003).

Epigenetic regulation seems to be a combinatorial effect of multiple types of modifications in DNA and histone proteins. These modifications, which may or may not induce alterations in DNA packaging, are able to synchronize the expression of hundreds or thousands of genes dispersed in large genomic segments and simultaneously modulate the expression of limited numbers of genes in the same or similar regions.

However, order can be created from disorder. Here we will focus on EM study by antibodies, *i.e.*, by the ChIP-EM approach.

EXPERIMENTAL PROCEDURE

Mapping of the genetic modification sites is made possible by using antibodies, each of which reacts to a specific type of modification in a specific location of the histone tails, to enrich the chromatin fragments containing the modification of interest (Zhao *et al.*, 2007). Sequencing of the DNA portion of the enriched chromatin fragments followed by mapping sequences against the appropriate reference genome, the genome-wide locations of the modification can be identified. This type of information can be correlated to other information for a comprehensive understanding of the cellular status (Figs. **1** and **2**).

Construction of ChIP-EM Sequencing Libraries

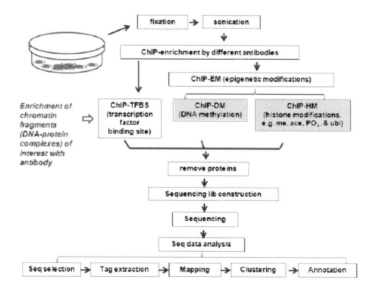

Fig. (1). Procedure for ChIP-EM (epigenetic modification) analysis.

Sequence Analysis of ChIP-EM Libraries

ChIP-HM (histone modifications) detected by paired-end ditag approach

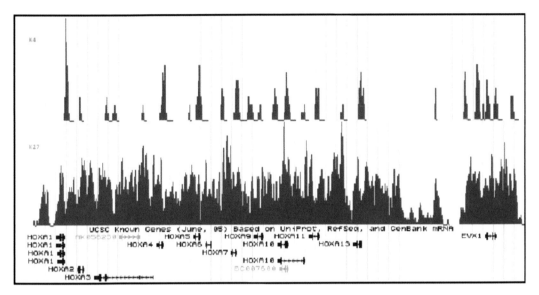

Fig. (2). ChIP-PET analysis of H3K4me3 and H3K27me3 in the homeobox domain.

REFERENCES

Arnaudo, A.M., Garcia, B.A. (2013). Proteomic characterization of novel histone post-translational modifications. *Epigenetics Chromatin,* *6*(1), 24.
[http://dx.doi.org/10.1186/1756-8935-6-24] [PMID: 23916056]

Cedar, H., Bergman, Y. (2009). Linking DNA methylation and histone modification: patterns and paradigms. *Nat. Rev. Genet., 10*(5), 295-304.
[http://dx.doi.org/10.1038/nrg2540] [PMID: 19308066]

Fischle, W., Wang, Y., Allis, C.D. (2003). Histone and chromatin cross-talk. *Curr. Opin. Cell Biol., 15*(2), 172-183.
[http://dx.doi.org/10.1016/S0955-0674(03)00013-9] [PMID: 12648673]

Gräff, J., Kim, D., Dobbin, M.M., Tsai, L.H. (2011). Epigenetic regulation of gene expression in physiological and pathological brain processes. *Physiol. Rev., 91*(2), 603-649.
[http://dx.doi.org/10.1152/physrev.00012.2010] [PMID: 21527733]

Hirabayashi, Y., Gotoh, Y. (2010). Epigenetic control of neural precursor cell fate during development. *Nat. Rev. Neurosci., 11*(6), 377-388.
[http://dx.doi.org/10.1038/nrn2810] [PMID: 20485363]

Kouzarides, T. (2007). Chromatin modifications and their function. *Cell, 128*(4), 693-705.
[http://dx.doi.org/10.1016/j.cell.2007.02.005] [PMID: 17320507]

Waddington, C.H. (2012). The epigenotype. 1942. *Int. J. Epidemiol., 41*(1), 10-13.

[http://dx.doi.org/10.1093/ije/dyr184] [PMID: 22186258]

Zhao, X.D., Han, X., Chew, J.L., Liu, J., Chiu, K.P., Choo, A., Orlov, Y.L., Sung, W.K., Shahab, A., Kuznetsov, V.A., Bourque, G., Oh, S., Ruan, Y., Ng, H.H., Wei, C.L. (2007). Whole-genome mapping of histone H3 Lys4 and 27 trimethylations reveals distinct genomic compartments in human embryonic stem cells. *Cell Stem Cell, 1*(3), 286-298.
[http://dx.doi.org/10.1016/j.stem.2007.08.004] [PMID: 18371363]

CHAPTER 14

MicroRNA Analysis

Abstract: MicroRNAs (miRNAs) negatively regulate mRNA species by binding to the 3' untranslated region (3' UTR) in mRNA through nucleotide complementarity which allows limited number of nucleotide mismatches to fine tune the target specificity and the degree of represion. Although miRNAs have been intensively studied for decades, most of their targets and functions remain unknown. Furthermore, many of the miRNAs that have been studied are known to target multiple mRNAs. These properties seriously impede the progress of miRNA analysis. Analysis of miRNAs normally relies on commercial kits for miRNA isolation and sequencing library preparation. This chapter will serve as a general introduction of miRNA analysis. Most of the experimental procedure and sequence data analysis discussed in this chapter can also be found in the paper entitled "global assessment of *Antrodia cinnamomea*-induced microRNA alterations in hepatocarcinoma cells" published in 2013.

Keywords: 3' UTR, MicroRNA, MiRNA, Non-coding RNA.

INTRODUCTION

miRNAs are small single-stranded non-coding RNAs of ~18–24 nucleotides that have been postulated to post-transcriptionally regulate up to 50% of genes in both plants and animals (Bushati and Cohen, 2007; Friedman *et al.*, 2009; Jovanovic and Hengartner, 2006; Kaikkonen *et al.*, 2011). The biogenesis of miRNA follows a distinct pathway. Similar to mRNAs, most miRNA genes are transcribed by RNA polymeras II to generate primary miRNA (pri-miRNA) which also contains a 5'cap and a 3' polyA. A complex of Drosha and DGCR/Pasha cleaves the pri-RNA to produce ~70nt hairpin-shaped precursor miRNA (pre-miRNA), which is subsequently transported by Exporin-5 to the cytoplasm (Bohnsack *et al.*, 2004; Du and Zamore, 2005; Yi *et al.*, 2003), where the pre-miRNA is cleaved by a complex of Dicer and TRBL/Loquacious, releasing the double stranded ~21nt (miRNA-miRNA* duplex) mature miRNA. In most cases, the miRNA* strand is degraded, whereas the 5' end of miRNA is incorporated into an RNA-induced silencing complex (RISC) with Argonaut proteins to regulate its target mRNA. Through binding to the 3'UTR of its target gene, a miRNA can either degrade the target mRNA or repress its translation (Bushati and Cohen, 2007; Catto *et al.*, 2011; Du and Zamore, 2005). We have recently reported the global downregulation of miRNA by *Antrodia cinnamomea* fungus (Chen *et al.*, 2013).

Kuo Ping Chiu

EXPERIMENTAL PROCEDURE

Part 1: Construction of Sequencing Libraries

Preparation of the starting material

First prepare a batch of total RNA, which is expected to contain a certain amount of small RNA/miRNA

1. Determine the quantity of total RNA and the quality of small RNA.
   The quantity of total RNAcan be determined by using RNA 6000 Nano Kit in NanoDrop, while the quality of small RNA can be evaluated by, for example, RIN (RNA Integrity Number) as estimated by software. The RIN value reflects the degree of RNA degradation. If RIN > 6, then continue to the next step; otherwise, prepare a fresh sample.

2. Calculate the *percentage* of small RNA in your total RNA sample.
   Quantify the amount of 10-40 miRNA using Small RNA Chip. (% of miRNA= (amount of 10-40 miRNA divided by the amount of total RNA)*100.)
   If miRNA% ≥ 0.5%, then skip this step. That is, you can use the total RNA directly without further isolation. Otherwise, enrich the miRNA (using a commercial kit).
   We strongly recommend to ignore the miRNA% and proceed with miRNA enrichment even if the percentage is higher than the threshold given in A above, because non-miRNA species may complicate downstream reactions.

3. Enrich small RNA
   If miRNA% < 0.1%, flashPAGE Fractionator together with the flashPAGE Cleanup Kit can be used, and the size distribution of the "purified" sample is expected to be within 10-40 nt. If miRNA% ranges between 0.1-0.5%, PureLink miRNA Isolation Kit can be used for enrichment, and the size distribution of the enriched sample is expected to be within 10-200 nt.

4. Evaluate both the quality and quantity of the enriched small RNA sample
   The quality and quantity can be estimated by Agilent 2100 Bioanalyzer or fragment analyzer (AATI). (% of miRNA= (amount of 10-40 miRNA divided by the total amount of enriched RNA sample)*100).

5. Determine the quantity of input

Type of input	Estimated conc. of miRNA w/ size between 10-40 nt (on Small RNA Chip)	RNA input
Total RNA	5 – 100 ng/mL	≤ 1 mg
Enriched small RNA (size within 10-200 nt	1 – 100 ng/ mL	≤ 1 mg
flashPAGE-enriched small RNA (size within 10-40 nt)	1– 100 ng/ mL	0.1 mg

Part 2: Construction of the Amplified Small RNA Library

The general procedure includes the following steps: 1) hybridization and ligation of small RNA molecules to reverse transcriptase (RT) adaptors; 2) reverse transcription; 3) purification of cDNA; 4) size selection of cDNA (*e.g.* by gel excision); 5) amplification of cDNA by PCR; 6) purification of PCR-amplified DNA; and 7) assessment of the yield and size distribution of the amplified DNA (*e.g.* using Fragment Analyzer™ (AATI)); and 8) sequencing library construction.

Part 3: miRNA and mRNA Data Processing and Analysis

This section aims to give the readers a general idea, and an example, regarding how the miRNA sequence data is processed and analyzed.

For each miRNA library, the qualified sequence reads will be retained for downstream analysis by meeting the criterion of a Phred Quality Score of QV $\geq$ 20, which is equivalent to 99% accuracy. To process four libraries (2 hr or 4 hr for each, treated or untreated) of sequences with lengths of 35 bp generated from a sequencer, we built an in-house pipeline using shell script combined with custom-made perl script. To avoid having to trim adapters by using duplicated datasets, each identical sequence was clustered and assigned a unique tag with the bp count of the sequence (*e.g.*, tag_100). Cutadapt (Martin, 2011) was used to trim ligated adapters with fewer than 3 mismatches. Then, polyN was cleaned and identical reads shorter than 16 bp and longer than 30 bp were removed to match the range adopted by miRBase using custom made perl script.

To identify known and novel miRNAs, we eliminated tRNA and rRNA sequences from the libraries by mapping sequence reads against Rfam (http://rfam.sanger.ac.uk/) and tRNA (http://lowelab.ucsc.edu/GtRNAdb/) using the Bowtie program (Langmead *et al.*, 2009). Sequence reads mapped to those databases without mismatch were removed from the libraries. In addition, repetitive sequences in the libraries that mapped to Repbase (http://www.girinst.org/repbase/) were also removed, after which miRBase (http://www.mirbase.org/) was employed to identify the known miRNAs from the libraries. The remaining sequence reads were mapped to the UCSC hg19 database and only sequence reads successfully mapped to the UCSC hg19 with fewer than 3 mismatches were considered to be novel miRNAs. In our miRNA profile, we only used detected known miRNA with at least 2 reads or more for subsequent analysis.

DISCLOSURE

Part of this chapter has been previously published in PLoS ONE 8(12): e82751. doi:10.1371/journal.pone.0082751. 2013

REFERENCES

Bohnsack, M.T., Czaplinski, K., Gorlich, D. (2004). Exportin 5 is a RanGTP-dependent dsRNA-binding protein that mediates nuclear export of pre-miRNAs. *RNA, 10*(2), 185-191.
[http://dx.doi.org/10.1261/rna.5167604] [PMID: 14730017]

Bushati, N., Cohen, S.M. (2007). microRNA functions. *Annu. Rev. Cell Dev. Biol., 23*, 175-205.
[http://dx.doi.org/10.1146/annurev.cellbio.23.090506.123406] [PMID: 17506695]

Catto, J.W., Alcaraz, A., Bjartell, A.S., De Vere White, R., Evans, C.P., Fussel, S., Hamdy, F.C., Kallioniemi, O., Mengual, L., Schlomm, T., Visakorpi, T. (2011). MicroRNA in prostate, bladder, and kidney cancer: a systematic review. *Eur. Urol., 59*(5), 671-681.
[http://dx.doi.org/10.1016/j.eururo.2011.01.044] [PMID: 21296484]

Chen, Y.J., Thang, M.W., Chan, Y.T., Huang, Y.F., Ma, N., Yu, A.L., Wu, C.Y., Hu, M.L., Chiu, K.P. (2013). Global assessment of *Antrodia cinnamomea*-induced microRNA alterations in hepatocarcinoma cells. *PLoS One, 8*(12), e82751.
[http://dx.doi.org/10.1371/journal.pone.0082751] [PMID: 24358224]

Du, T., Zamore, P.D. (2005). microPrimer: the biogenesis and function of microRNA. *Development, 132*(21), 4645-4652.
[http://dx.doi.org/10.1242/dev.02070] [PMID: 16224044]

Friedman, R.C., Farh, K.K., Burge, C.B., Bartel, D.P. (2009). Most mammalian mRNAs are conserved targets of microRNAs. *Genome Res., 19*(1), 92-105.
[http://dx.doi.org/10.1101/gr.082701.108] [PMID: 18955434]

Jovanovic, M., Hengartner, M.O. (2006). miRNAs and apoptosis: RNAs to die for. *Oncogene, 25*(46), 6176-6187.
[http://dx.doi.org/10.1038/sj.onc.1209912] [PMID: 17028597]

Kaikkonen, M.U., Lam, M.T., Glass, C.K. (2011). Non-coding RNAs as regulators of gene expression and epigenetics. *Cardiovasc. Res., 90*(3), 430-440.
[http://dx.doi.org/10.1093/cvr/cvr097] [PMID: 21558279]

Langmead, B., Trapnell, C., Pop, M., Salzberg, S.L. (2009). Ultrafast and memory-efficient alignment of short DNA sequences to the human genome. *Genome Biol., 10*(3), R25.
[http://dx.doi.org/10.1186/gb-2009-10-3-r25] [PMID: 19261174]

Martin, M. (2011). Cutadapt removes adapter sequences from high-throughput sequencing reads. *EMBnet J., 17*, 10-12.
[http://dx.doi.org/10.14806/ej.17.1.200]

Yi, R., Qin, Y., Macara, I.G., Cullen, B.R. (2003). Exportin-5 mediates the nuclear export of pre-microRNAs and short hairpin RNAs. *Genes Dev., 17*(24), 3011-3016.
[http://dx.doi.org/10.1101/gad.1158803] [PMID: 14681208]

CHAPTER 15

Application of NGS in the Study of Sequence Diversity in Immune Repertoire

Abstract: During evolution, the immune system evolved as a defense mechanism to protect organisms against pathogens. Since pathogens in the environment are extremely diverse and unpredictable, strategies taken by the immune system have to be highly diversified in order to mount an effective response. At the molecular level, the sequence diversity present in the variable regions of antibody-coding and TCR-coding genomic sequences is eventually reflected in the amino acid sequences of their encoded proteins as seen in the circulation system and on the surface of immune cells. At the cellular level, B cells, T cells, dendritic cells, and many other immune cells have to interact coordinately with one another in order to foster the maturation of an immune response (*e.g.*, affinity maturation) against pathogenic attack. With the advent of NGS technologies, the complexity of the immune system can now be studied in greater detail.

Keywords: Phage display, Single chain variable fragment, ScFv, Single domain antibody, SdAb, TCR, VDJ recombination.

INTRODUCTION

With next-generation sequencing, we are now able to study the immune repertoire (*e.g.*, VDJ recombination events during B cell maturation) by high throughput sequencing of genomic sequences or mRNAs from various stages of B-cell or T-cell development. The amino acid sequences of their corresponding proteins can be deduced from their corresponding nucleotide sequences. Previously most studies focused on understanding immune processes by deducing function from structure (structure -> function). We now can add 'sequence' to the upstream of the process (sequence-> structure-> function). Details are illustrated in the following sections.

PART I. SEQUENCING THE IMMUNE REPERTOIRE

A. Characterization of a Natural Antibody Repertoire

Antibody repertoires are highly plastic and can be directed to produce antibodies with broad chemical diversity and extremely high selectivity. The work published by Weinstein *et al.*, in 2009 represents a great example to demonstrate how an immune repertoire can be analyzed comprehensively and thoroughly by deep sequencing

(Weinstein *et al.*, 2009). In the work, the authors studied the variable domain of the antibody heavy chain and analyzed the VDJ usage using mRNA samples isolated from zebrafish. In principle, similar approach can be applied to the study of TCR variable regions. Moreover, by using the genomic DNA, instead of mRNA, it can be applied to study the recombination events inside the nucleus. The experimental procedure adopted by Weinstein *et al.* is summarized below.

Experimental Procedure

First, multiple wild type zebrafish were collected and each fish was homogenized in the presence of TRIzol. Total RNA was then extracted from the fish and mRNA was subsequently isolated from the total RNA. Then, cDNA libraries were synthesized using superScript™ III reverse transcriptase. PCR amplification using 27 forward primers located within the consensus leader sequences for 39 functional V gene segments together with reverse primers located within the first 100bp of the IgM and IgZ constant domain were conducted to capture the entire complementarity-determining region 3 (CDR3), which should contain the vast majority of antibody diversity. PCR amplified fragments were then subjected to next-generation sequencing using 454 FLX.

The sequencer output a total of 640 million bases from 14 zebrafish, equivalent to 28,000-112,000 useful sequence reads per zebrafish. By sequence alignment to a reference genome, sequence reads were mapped to V and J segments in the genome (success rate reached 99.6%; failures were mainly caused by similarity in the V segment). Alignment to the D segment was determined within the VJ region to all reads using a clustering algorithm, The success rate was 69.6%, and many of the unassignable reads had D segments mostly deleted.

Their work resulted in a number of discoveries. For example, they found that 50-86% of all possible VDJ combinations were used and that zebrafish shared a similar frequency distribution of VDJ usage. Moreover, there was a correlation of VDJ patterns between individuals. They also demonstrated an evidence of convergence, as indicated by the fact that different individuals may make the same antibody.

B. Sequencing scFv and sdAb for Therapeutics

Recent immune engineering using synthesized DNA sequences, which may contain degenerate and constant regions, has been able to create chimeric antibodies for therapeutic purposes. This strategy opened up a new dimension for the usage of next-generation sequencing (Chang *et al.*, 2014; Hsu *et al.*, 2014).

For example, a single chain variable fragment (scFv) is a simplified fusion antibody produced by directly linking the variable regions of a heavy chain and a light chain through a "linker peptide" of ~25 aa in length. The order of V_H and V_L is interchangeable, making it either N-(V$_H$)-(linker peptide)-(V$_L$)-C or N-(V$_L$)-(linker peptide)-(V$_H$)-C in structure. Since both the variable regions of the heavy chain and the light chain are present in its structure, an scFv can bind its antigen with high specificity. Moreover, the structure and function of their coding sequence can be easily tested using phage display libraries. As such, this scFv provides a convenient approach for quick Ab production and testing for therapeutics.

Selection of high-affinity variable sequences using phage display screening

Phage display screening is a powerful approach for selecting high-affinity variable sequences from an immune repertoire for sequencing. A general workflow is shown in Fig. **(1)**.

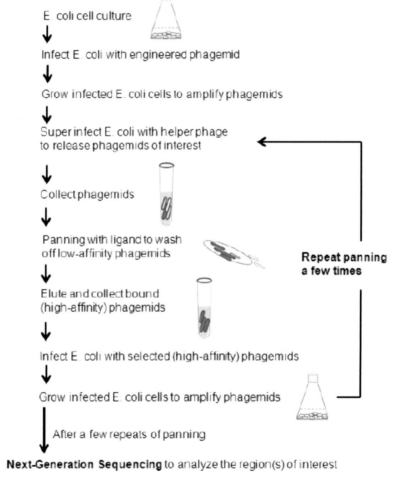

Fig. (1). Phage display screening for high-affinity phagemids. Phage display screening enriches high-affinity (or low-affinity if the counterpart is desired) phagemids. The procedure consists of a few transductions, growth of bacterial cultures and a series of panning used to separate low-affinity phagemids from its high-affinity counterpart.

PCR using primers located in the constant regions can be used to amplify the desired variable region(s) in phage display libraries, in the genomes of B or T cells (immature or mature) or the transcripts of B or T cells (immature or mature), or in a scFv or sdAb engineered library. The amplified regions are then subjected to NGS analysis.

For sequencing, a number of issues need to be considered, including 1) selection of 454 or Illumina sequencer; 2) determination of the optimal sequence length needed for collecting the desired information; and 3) whether to conduct paired-end sequencing, fragment sequencing, or paired-end ditag (mate-pair) sequencing.

After sequencing, sequence alignment (mapping) is essential for sequence data analysis. Normally only the variable regions are analyzed. One may want to identify the predominant sequence species for a particular antigen and the most variable region in the sequence data. Also, one may want to compare the nucleotide sequences with their corresponding amino acid sequences for transcriptome-proteome cross comparison.

C. Single Domain Antibody (sdAb)

Single domain antibody, or nanobody, is another example of simplified artificial antibody. It consists of a single monomeric variable antibody domain, either from a heavy chain or a light chain. Similar to a whole Ab and scFv, an sdAb is able to bind its antigen with specificity. This ingenious engineering procedure must be accompanied by a display system and *in vitro* testing.

Advantages and disadvantages of scFv and sdAb

There are advantages of scFv and sdAb. These include 1) affinity maturation can be accomplished either by error-prone PCR (Fleishman *et al.*, 2011), or by other mutagenesis approaches; 2) these technologies are able to quickly design artificial antibodies; 3) such simplified antibodies can be produced by, and expressed on the surfaces of, microbes such as phages, bacteria and yeast; 4) in the phage display cycle (bind-> wash-> elute-> amplification), the affinity strength can be adjusted by the cycle number of panning or pH value; 5) these antibodies are suitable for analysis of a molecular population; 6) there is no need to depend solely on traditional animals, such as the horse, mouse, or rabbit, for antibody production; and 7) sequence-function testing can be completed quickly (*e.g.* by phage display screening).

On the other hand, there are also disadvantages which may hinder their applications. These include 1) the specificity cannot be as high as that for whole antibodies; 2) the turnover rate of engineered scFv and sdAb molecules is rapid, making the resultant immunity short and not sustainable; 3) immunity provided by scFv or sdAb cannot be sustained by immune B and T cells; and 4) side effects have to be evaluated.

PART II. APPLICATION OF NGS IN VACCINE DEVELOPMENT

Vaccine development involves a number of important issues. These include 1) regulation; 2) identification of suitable antigens, adjuvants, and methods of delivery; 3) technical hurdles; 4) manufacturing capacity; 5) the need for lengthy clinical trials (10-15 yrs); and 6) the costs and benefits of public and private participation.

Vaccine development is a lengthy process which can be divided into a few stages. During the initial stage of vaccine development, a working hypothesis can be very helpful in guiding our thinking and experimentation. Frequently we need to make a best guess, or a bold guess, based on empirical data for designing a vaccine. Contamination in the vaccine may cause death of the recipient. When preparing a vaccine, the possibility of having comtanination in the vaccine has to be minimized and carefully monitored. It is also important to record the strength and the types of responses, together with toxicity and side effects, provoked by the vaccination when testing a vaccine in an animal model.

Applications of NGS Technologies in Vaccine Development and Pathogen Control

There are ample applications of sequencing in pathogen control. The best known applications include viral genome sequencing, detection of susceptibility-associated viral genome variations, and the study of virus-host interactions (*e.g.* SARS genome sequencing in early 2003). The advent of NGS technologies not only facilitate these applications, but also create more applications for biological study. With unprecedented advantageous features such as high yield, high speed and low cost, NGS has been routinely applied to the study of cancer-associated genetic and epigenetic mutations, and the detection of low-abundance drug-resistant mutations in the microbial genomes and disease- or infection-associated susceptibility genes. These applications directly or indirectly facilitate vaccine development, which can be examplified by the cases shown below.

The first case is related to the vaccine against rotavirus, which is the most common cause of severe diarrhea among infants and young children. GlaxoSmithKine developed a live-attenuated rotavirus vaccine, called Rotarix. Deep sequencing by NGS revealed that Rotarix vaccine was contaminated with porcine circovirus 1 DNA (Victoria *et al.*, 2010). Consequently, the use of Rotarix in the US was suspended. In this case, sequencing technology demonstrated itself as a robust tool able to ensure the quality of a vaccine.

The second case is related to the characterization of BCG attenuated vaccines. *Mycobacterium bovis* bacillus Calmette-Guerin (BCG) is the only vaccine available against TB (Gomes *et al.*, 2011). This attenuated strain was derived from *M. bovis* isolate after 230 serial passages *in vitro* at the Institute Pasteur and then distributed to laboratories worldwide for vaccination against TB. Thus, the BCG vaccines produced worldwide represent a heterogeneous family of daughter strains. Genome sequencing is needed to identify sequence variations and to prevent contamination.

For sequencing a microbial genome, a non-NGS sequencer may be sufficient. For sequencing multiple microbial genomes, barcodes can significantly facilitate the process.

REFERENCES

Chang, H.J., Jian, J.W., Hsu, H.J., Lee, Y.C., Chen, H.S., You, J.J., Hou, S.C., Shao, C.Y., Chen, Y.J., Chiu, K.P., Peng, H.P., Lee, K.H., Yang, A.S. (2014). Loop-sequence features and stability determinants in antibody variable domains by high-throughput experiments. *Structure, 22*(1), 9-21.
[http://dx.doi.org/10.1016/j.str.2013.10.005] [PMID: 24268648]

Fleishman, S.J., Whitehead, T.A., Ekiert, D.C., Dreyfus, C., Corn, J.E., Strauch, E.M., Wilson, I.A., Baker, D. (2011). Computational design of proteins targeting the conserved stem region of influenza hemagglutinin. *Science, 332*(6031), 816-821.
[http://dx.doi.org/10.1126/science.1202617] [PMID: 21566186]

Gomes, L.H., Otto, T.D., Vasconcellos, E.A., Ferrão, P.M., Maia, R.M., Moreira, A.S., Ferreira, M.A., Castello-Branco, L.R., Degrave, W.M., Mendonça-Lima, L. (2011). Genome sequence of Mycobacterium bovis BCG Moreau, the Brazilian vaccine strain against tuberculosis. *J. Bacteriol., 193*(19), 5600-5601.
[http://dx.doi.org/10.1128/JB.05827-11] [PMID: 21914899]

Hsu, H.J., Lee, K.H., Jian, J.W., Chang, H.J., Yu, C.M., Lee, Y.C., Chen, I.C., Peng, H.P., Wu, C.Y., Huang, Y.F., Shao, C.Y., Chiu, K.P., Yang, A.S. (2014). Antibody variable domain interface and framework sequence requirements for stability and function by high-throughput experiments. *Structure, 22*(1), 22-34.
[http://dx.doi.org/10.1016/j.str.2013.10.006] [PMID: 24268647]

Victoria, J.G., Wang, C., Jones, M.S., Jaing, C., McLoughlin, K., Gardner, S., Delwart, E.L. (2010). Viral nucleic acids in live-attenuated vaccines: detection of minority variants and an adventitious virus. *J. Virol., 84*(12), 6033-6040.
[http://dx.doi.org/10.1128/JVI.02690-09] [PMID: 20375174]

Weinstein, J.A., Jiang, N., White, R.A., III, Fisher, D.S., Quake, S.R. (2009). High-throughput sequencing of the zebrafish antibody repertoire. *Science, 324*(5928), 807-810.
[http://dx.doi.org/10.1126/science.1170020] [PMID: 19423829]

Part 3

Introduction to Analytical Tools

CHAPTER 16

Galaxy Pipeline for Transcriptome Library Analysis

Abstract: Next Generation Sequencing (NGS) provides researchers with an unprecedented opportunity to produce a large volume of DNA sequences quickly, and is one of the fundamental methods for high-throughput genomic studies. Currently, the most widely-used NGS platforms are Illumina, Roche 454 and SB SOLiD. These platforms differ in the chemistry used in the sequencing process and the length of sequencing read generated. Each platform has its own strengths and weaknesses. In particular, the required length of the sequence read to be generated plays an important role when designing an experiment. For example, a longer read length would be needed in the assembly of a novel genome, while throughput-maximizing PED-based techniques would be better-suited when shorter reads will suffice.

Keywords: ChIP-Seq, Galaxy, RNA-Seq.

INTRODUCTION

Previous chapters have provided readers with the basic concepts behind various NGS platforms and their applications. In this chapter, we present tools that are widely used for data analysis. Over the years, a large number of tools have been created for analyzing NGS data. Many of these tools require creating local databases, and familiarity with the Unix/Linux operating system, which poses a huge challenge to non-computational scientists, is needed. Galaxy was designed to overcome these issues by providing an open, web-based platform that integrates various databases and tools together with a simple interface for performing NGS data analysis.

GALAXY INTERFACE

Galaxy is a popular pipeline for RNA-Seq (Trapnell *et al.*, 2012). It is available from http://usegalaxy.org. The interface is simple and intuitive, as shown in Fig. (**1**). This is the Analyze Data window where users can perform data analysis. The interface is organized into four main sections: the top Menu bar, Tools panel (left column), Operation panel (middle column), and History panel (right column). The Tools panel provides the users a list of available algorithms that are useful for processing and analyzing NGS data. The Operation panel is used as the input interface for tools and information display. The History panel keeps track of all operations, parameters, and input and output datasets.

Kuo Ping Chiu

Fig. (1). Galaxy interface is organized into four sections: Menu bar (top), Tools panel (left), Operation panel (middle), and History panel (right). The menu bar allows users to switch between various functional interfaces. The Tools panel contains all the available data analysis integrated into Galaxy. The middle panel is used as the input interface for an algorithm and to display information or results. Pictured here is the Upload File interface. The History panel keeps track of the datasets and operations. There are three datasets in the current history. Datasets 1 and 2 are the input datasets, and dataset 7 is the result of applying a workflow program to the two input datasets.

There are a large number of tools integrated into Galaxy. These tools are grouped into various sections, according to their functions. For example, the Get Data section contains a list of tools for getting data into Galaxy. These tools allow users to upload data from their computer or retrieve information from public resources such as the UCSC Table Browser and the BioMart server.

In the following sections, we describe the functionality of Galaxy for analyzing ChIP-Seq data. ChIP-Seq is a method used to study protein-DNA interactions, which it does by replacing the microarray ("chip") used in the ChIP-chip technique with sequencing ("Seq"). These methods have been widely used to identify the DNA-binding sites of transcription factors and the locations of histone modification.

FASTQ FORMAT

Many sequencing platforms can be used for sequencing. While they vary in the ways they encode the quality scores, as exemplified below, these sequencing platforms produce data in the standard Fastq format. And, unlike the old Fasta format, the scores of all nucleotides in a read are also included. Each read is represented in four lines (Fig. **2a**): 1) the header line, which begins with an '@' character and is followed by a sequence identifier; 2) the raw sequence letters; 3) the third line begins with a '+' character and is optionally followed by the same sequence identifier; and 4) the fourth line gives the quality scores.

```
a
@HWI-ST_0216:7:1101:1742:2161#CGATGG/1
AGCCGGGACCCCGCCGATCACACCGGAGATGGTCCGTCGCGCGCTCGATGAGGACTGACGCGGGTGGCGCTGCTCGACGTCAACGCATTGGTCGCGCTGG
+HWI-ST_0216:7:1101:1742:2161#CGATGG/1
___cccc^beeZc^fa[_U^Z^eZ`eggUbHHM\\`a_[a^aaWVOY_b]]bbb_X^`GTZaaaTZa[]aaaOSYWLOX]OGT]_aX^ab_]]]O[][aa
@HWI-ST_0216:7:1101:1679:2241#ATGGCA/1
GCACCAACATTGATGAGCTGCTACTGGAACAACCCGGAGGCCACCGCGGAGGCGTTCGCAGGCGGCTGGTTCCATTCTGGGGATCTGGTTCGTATGGACT
+HWI-ST_0216:7:1101:1679:2241#ATGGCA/1
_^_eeeee^cgaegd`ehgg`fhegghbaffddPbec`egddgfefhfgc`^^a^^W^aV_cZacaTZabcbbbc_b`]`b_a^_bb_bba[[b[X_`b

b
SSSSSSSSSSSSSSSSSSSSSSSSSSSSSSSSSSSSSSSSSSSSSSSSSSSS....................................................
...........................XXXXXXXXXXXXXXXXXXXXXXXXXXXXXXXXXXXXXXXXXXXXXX....................................
..........................................IIIIIIIIIIIIIIIIIIIIIIIIIIIIIIIIIIIIIIIIIIIIIIIIIIIII....................................
...........................................JJJJJJJJJJJJJJJJJJJJJJJJJJJJJJJJJJJJJJJJJJJJJJJJ.....................................
..LLLLLLLLLLLLLLLLLLLLLLLLLLLLLLLLLLLLLLLLLLLLLLLLLL...................................................
!"#$%&'()*+,-./0123456789:;<=>?@ABCDEFGHIJKLMNOPQRSTUVWXYZ[\]^_`abcdefghijklmnopqrstuvwxyz{|}~
|                          |    |    |                                          |                   |
33                         59   64   73                                         104                 126
0.....................26...31.......40
                       -5....0........9...............................40
                       0........9...............................40
                               3......9............................40
0.2...................26...31........41

S - Sanger         Phred+33,   raw reads typically (0, 40)
X - Solexa         Solexa+64,  raw reads typically (-5, 40)
I - Illumina 1.3+  Phred+64,   raw reads typically (0, 40)
J - Illumina 1.5+  Phred+64,   raw reads typically (3, 40)
    with 0=unused, 1=unused, 2=Read Segment Quality Control Indicator (bold)
    (Note: See discussion above).
L - Illumina 1.8+  Phred+33,   raw reads typically (0, 41)
```

Fig. (2). An example of Fastq format (**a**). Each sequence is represented in four lines, sequence header, sequence, quality header, and quality line. Various schemes can be used to encode quality scores (**b**). The numeric scores are converted to the corresponding ASCII code.

As noted, each platform uses different schemes to encode the quality scores. The initial scheme was introduced by the Sanger Institute; it encodes a Phred quality score from 0 to 93 using ASCII 33 to 126. The currently available encoding schemes are shown in Fig. (**2b**). Please note that the analysis modules in Galaxy require Sanger quality scores.

DATASET

Galaxy uses the concept of "dataset" to represent the data and applies operations on these datasets. The datasets could be files uploaded from an individual's computer, information extracted from public resources, or results of prior analyses. Each dataset has associated metadata, which provides information such as the name or description of the dataset, the datatype, associated reference genome, and input parameters. The History panel keeps track of all the datasets and can be saved for subsequent analysis or reference. Galaxy provides sample datasets for users to help them master its functionality. These shared datasets can be accessed by clicking "Shared Data" in the Tools panel, and then selecting the "Data Libraries". Currently, there are more than 30 datasets available to users. For example, the "Sample NGS Datasets" contain sequencing samples from three major NGS platforms, Illumina, Roche 454, and SOLiD.

EXERCISE

In the following exercises, we will use the datasets from the "ChIP-Seq Mouse Example" library to exemplify the workflow for analyzing ChIP-Seq data using Galaxy. The original data files from the ENCODE project were too large and have been scaled down to contain only data aligned to chromosome 19 of the mouse genome. Two files are available. One is the data from a mouse ChIP-Seq experiment and the other from its corresponding control experiment.

USER ACCOUNT

While you can start using the Galaxy system immediately to analyze your data, it is advisable to create a user account in order to take full advantages of the system. With a user account, you can save your work, share it with others and have access to the Workflow tool. To create an account, click "User" on the top menu bar, select "Register", and fill in the details.

DOWNLOAD THE DATA

The datasets we will use for the exercises in this chapter are derived from a ChIP-Seq experiment treated with CTCF-targeting antibody and its control. CTCF is a transcription factor that has been experimentally associated with cancer tumors. These datasets are a scaled-down version from the original data, containing only reads mapped to chromosome 19 of the mouse genome (mm9). Follow the steps below to download the datasets.

1. Open the Galaxy server by entering http://usegalaxy.org in your web browser, and logon with your account.
2. Click on "Shared Data" in the top menu, and choose "Data Libraries".

3. Click on "ChIP-Seq Mouse Example".

4. Check the "Mouse ChIP-Seq example Control Data, chr19, mm9" and "Mouse ChIP-Seq Example Experimental Data, chr19, mm9" datasets.

5. Set the pull-down menu for "For selected datasets:" to "Download as a .tar.gz file" and click the "Go" button.

6. Save the file "ChIP-Seq_Mouse_Example_files.tar".

7. Untar the file. This will give you a folder named "ChIP-Seq Mouse Example". This folder contains two files, "Mouse ChIP-Seq example Control Data, chr19, mm9" and "Mouse ChIP-Seq Example Experimental Data, chr19, mm9".

8. Compress the two files using the gzip (or zip) program. This can speed up the upload progress. The compressed files will automatically be decompressed.

UPLOAD THE DATA

There are several ways to get files or datasets into the Galaxy server. You can upload your files directly using a web browser. However, for a large file, you are recommended to first upload the file to the Galaxy FTP server using your account information and select it from "Files uploaded *via* FTP" later (Fig. 1). In addition, Galaxy can also retrieve a file if it is publicly accessible on the internet. Follow these steps to upload the data.

1. Click on "Analyze Data" in the top menu bar to go back to the analysis interface.

2. Click the "Get Data" at the top of the Tools panel, and select "Upload File".

3. This will load the upload file tool in the Operation panel (Fig. **3a**).
   a. Set "File Format:" to "fastq".
   b. Use the Browse button to set "File:" to "Mouse ChIP-Seq Example Experimental Data, chr19, mm9.gz".
   c. Set "Genome:" to "Mouse July 2007 (NCBI37/mm9) (mm9)".
   d. Click "Execute" to upload the file. A new record will be created in the History panel. Each record is associated with a color to indicate the stage of completion. When a job is queued, the record appears in gray. When the file is being uploaded, the color changes to blue color. Once the file is completely uploaded, the color turns to green (successful) or red (error).
   e. Fig. (**3b**) shows the way the panel will work when a file is completely uploaded. Click the "eye" icon to display the sample data in the middle panel and "pencil" icon to edit the attributes (metadata) of the dataset. The "X" icon will remove the data from the history. Clicking on the item's name will display a preview window (Fig. **3c**). You can save or view details about this dataset by selecting the "save" or "information" icons. Use the "rerun" icon to rerun the job using the same settings (in case of an error execution) or with modifications.

4. Repeat step 3 to upload the remaining file, "Mouse ChIP-Seq example Control Data, chr19, mm9.gz".

5. Click on "Unnamed history" at the top of the History panel. Type in "Mouse ChIP-Seq

Example" and hit "Enter" to save.

Note: Galaxy also provides an easy way to load datasets from "Shared Data". Instead of downloading the files, you can set the pull-down menu for "For selected datasets:" to "Import to current history", and click the "Go" button. Click "Analyze Data" on the top menu bar and both datasets should be added to the History panel.

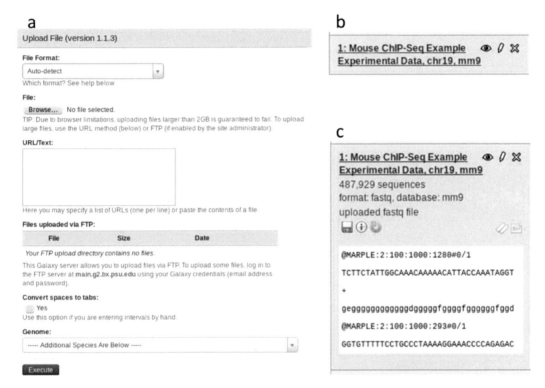

Fig. (3). The input interface for the upload file tool **(a)**. Galaxy accepts files uploaded from a local computer or FTP as well as from a file"s public URL. Uploading a large file directly is generally not recommended. A dataset that has been uploaded successfully is shown in green **(b)**. Colors are used to indicate the state of an execution, success (green), error (red), uploading (blue), and in progress (yellow). The eye icon displays the dataset in the Operation panel. Click on the pencil icon to edit the dataset"s attributes and the X icon to remove the dataset from the current history. Clicking on the name of a dataset will open the preview window **(c)**. The preview window gives a basic description of the dataset and sample data. Use the save icon to save the dataset locally and the information icon to view detailed information about the dataset. The rerun icon can be used to rerun a task with the same setting or modifications. This is very useful when an execution failed or when repeating the same task on other datasets.

FASTQ GROOMER

As mentioned earlier, Galaxy only accepts Sanger Fastq files. The datasets uploaded in a previous exercise were generated using the Illumina platform. Fortunately, however, Galaxy provides a tool to convert different Fastq variants to the "Sanger" Fastq format. Follow the steps below to convert file when necessary.

1. Click on "NGS: QC and manipulation" in the Tools panel to expand the list of available tools.
2. Under "ILLUMINA DATA", select "FASTQ Groomer".
3. Set "File to groom:" to "1: Mouse ChIP-Seq Example Experimental Data, chr19, mm9".
4. Confirm that "Input FASTQ quality scores type:" is set to "Sanger".
5. Click "Execute".
6. Once the job is completed, click on the pencil icon to change the name of the newly generated dataset. The default name was set to "FASTQ Groomer on data 1" automatically, which is not very informative. Change the name to "Mouse ChIP-Seq Example Experimental Data, chr19, mm9 (Groomed)". Change other attributes to suit your needs. Click "Save" when you complete the editing. The rerun tool can be used to repeat the same task on other datasets. This is very helpful in simplifying the task when a tool requires substantial parameter tuning.
7. The second dataset needs to be groomed as well. Click the "Mouse ChIP-Seq Example Experimental Data, chr19, mm9 (Groomed)" to expand the dataset, and click the rerun icon. The middle panel will display the same settings as for the previous run. Change the "File to groom:" to "2: Mouse ChIP-Seq example Control Data, chr19, mm9" and click "Execute".
8. Change the name of the newly generated dataset to "Mouse ChIP-Seq example Control Data, chr19, mm9 (Groomed)".

QUALITY CONTROL

Although all sequencing companies are constantly upgrading their technologies to improve sequencing quality, errors are inevitable. Sequencing quality tends to degrade toward the end of a read. Fig. (**4a**) shows how the read quality drops drastically after base 75. Given that the sequence length for this library is 100 bases, it is easy to see that as much as a quarter of this read is not reliable towards the end. These bases should not be used in mapping or downstream analysis. It is important to check the quality of your data before beginning your analysis. Galaxy provides a number of tools to ascertain the overall quality of your library. Here we will use the FastQC tool.

1. Click on "NGS: QC and manipulation" in the Tools panel to expand the list of available tools.
2. Scroll down to the bottom of the list, and select "FastQC: Read QC".
3. Set "Short read data from your current history:" to "3: Mouse ChIP-Seq Example Experimental Data, chr19, mm9 (Groomed)".
4. Click "Execute".
5. Once the job is completed, click the "eye" icon to view the result. The "Per base sequence quality" is shown in Fig. (**4b**). Overall, the quality seems to be quite good.
6. Apply the above steps to "4: Mouse ChIP-Seq example Control Data, chr19, mm9

(Groomed)". Again, the quality seems to be fine (Fig. **4c**).

Galaxy provides many tools to manipulate the Fastq files. These tools are listed in the "NGS: QC and manipulation" section. As mentioned above, the overall quality of reads tends to decrease toward the end of the sequence. It may be useful to trim the reads before engaging in a downstream analysis. This can be done by using the "FastQ Trimmer" in the "GENERIC FASTQ MANIPULATION" section. This tool trims 3' and 5' ends of every read in a dataset. It is also possible to filter out low quality reads from a dataset. The "Filter FastQ" tool allows users to remove reads that contain one or more low-quality bases. More complex Fastq manipulation tools are available under "FASTX-TOOLKIT FOR FASTQ DATA".

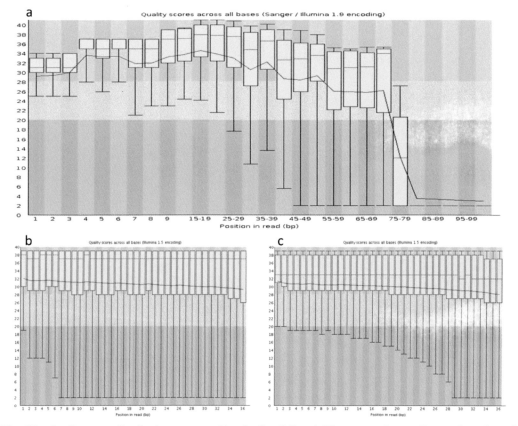

Fig. (4). Quality score distribution generated by the FastQC tool. The sequencing quality tends to degrade toward the end of reads. The first sample **(a)** shows that the read quality dropped drastically after base 75 and should not be used for downstream analysis. The other two examples **(b and c)** are the quality scores for the sample datasets used in this chapter. These two datasets are scaled-down versions of the original data and contains the reads that mapped to chromosome 19 of mouse genome (mm9).

MAPPING READS TO THE REFERENCE GENOME

The sequence reads represent the DNA/RNA fragments captured from the genome under a specific condition. Many tools have been designed to align these fragments back to the reference genome in order to identify their origins. These tools are mostly available as command-line programs, which increase the hurdles for biologists to fully utilize them. Again, it is fortunately that Galaxy integrates many of these algorithms within a single, simple framework, which can be accessed from the "NGS: Mapping" section, as specified below.

1. Click on "NGS: Mapping" in the Tools panel to expand the list of available mapping algorithms.
2. Select "Map with Bowtie for Illumina"
3. Set "Select a reference genome:" to "Mouse (Mus musculus): mm9 Canonical Male". In the search box, type "mm9" to perform a quick search. The "Canonical Male" means the reference genome contains all the autosomes, both sex chromosomes, and the mitochondrial genome. Unmapped contigs or scaffolds are not included.
4. Set "FASTQ file:" to "3: Mouse ChIP-Seq Example Experimental Data, chr19, mm9 (Groomed)".
5. Change "Bowtie settings to use:" to "Full parameter list" to display all available options. The "Commonly used" option might not be suitable for all situations.
6. Set "Maximum permitted total of quality values at mismatched read positions (-e):" to "80".
7. Set "Whether or not to try as hard as possible to find valid alignments when they exist (-y):" to "Try hard".
8. Check the "Suppress the header in the output SAM file:" option.
9. Click "Execute". Rename the dataset to "Mouse ChIP-Seq Example Experimental Data, chr19, mm9 (SAM)"
10. Repeat the above step by replacing the "FASTQ file:" to "4: Mouse ChIP-Seq example Control Data, chr19, mm9 (Groomed)". Use the rerun tool to save you from having to select the same settings again. Rename the dataset to "Mouse ChIP-Seq example Control Data, chr19, mm9 (SAM)".

IDENTIFYING BINDING SITES

Aligning the sequence reads back to the reference genome reveals the locations of the DNA/RNA fragments captured. In our example, the genomic locations can be used to predict the binding sites of the CTCF transcription factor, which binds to several thousand locations across the genome and functions as either a repressor or activator. In the following exercise, we will use the MACS (Model-based Analysis of ChIP-Seq) algorithm to identify the potential binding regions.

1. Click on "NGS: Peak Calling" to show the available tools.
2. Select "MACS".
3. Set the "ChIP-Seq Tag File:" to "7: Mouse ChIP-Seq Example Experimental Data,

chr19, mm9 (SAM)".

4. Set the "ChIP-Seq Control File:" to "8: Mouse ChIP-Seq example Control Data, chr19, mm9 (SAM)".

5. Set "Effective genome size:" to "1.87e+9". This is roughly the size of the mouse genome.

6. Set "Tag size:" to "36".

7. Set "Select the regions with MFOLD high-confidence enrichment ratio against background to build model:" to "32".

8. Check the box for "Parse xls files into distinct interval files". This will generate two additional interval files for peaks and negative peaks that might be useful for downstream analysis.

9. Set "Save shifted raw tag count at every bp into a wiggle file:" to "Save", and then set the "Resolution for saving wiggle files:" to "1". This will generate two WIG files for visualization purpose, "treatment" and "control".

10. Click "Execute".

11. Once the execution is completed, six new datasets will be listed in the History panel. The two standard output files are peaks listed in bed format and an html report. The other four are optional files generated by steps 8 and 9.

12. Click on the "pencil" icon on the bed file "9: MACS on data 8 and data 7 (peaks: bed)" to edit its attributes.

   a. Change the name to "CTCF Peaks, chr19, mm9 (BED)".

   b. Set "Score column for visualization:" to "5".

   c. Click "Save".

13. Visualize the binding sites in the UCSC browser.

   a. Click "9: CTCF Peaks, chr19, mm9 (BED)" to expand the preview panel. It shows that 720 regions have been identified.

   b. Click the link after "display at UCSC main". This will open the UCSC browser in another window, with the peaks given in the top track (Fig. 5).

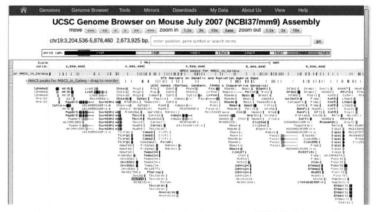

Fig. (5). Galaxy provides an easy way to visualize the data on the UCSC Genome Browser. The predicted binding sites from the MACS algorithm is shown on the first track and highlighted in green.

DOWNSTREAM ANALYSIS

We have identified 720 potential CTCF binding sites on chromosome 19, mm9. The next step is to ascertain how many of these peaks occur in the upstream regions of the known genes. Traditionally, this would require investigators to download the genomic data from public resources such as the UCSC Genome Browser, extract and process the genes, and compare with the binding sites. This process requires a number of tools and skills. Galaxy provides an easy way to perform these tasks using only a web browser.

Getting the genes from UCSC Table Browser:

1. Click "Get Data" to expand the list and select "UCSC Main". The UCSC Table Browser should appear in the middle panel (Fig. **6**). The History panel is temporary hidden for this operation. Click on the "<" icon in the bottom right corner to unhide it.
2. Set "clade:" to "Mammal".
3. Set "genome:" to "Mouse".
4. Set "assembly:" to "July 2007 (NCBI37/mm9)"
5. Set "group:" to "Genes and Gene Prediction Tracks".
6. Set "track:" to "RefSeq Genes".
7. Set "table:" to "refGene".
8. Set "region:" to "position" and enter "chr19" in the text field.
9. Set "output format:" to "BED – browser extensible data".
10. Make sure that the "Send output to Galaxy" box is checked.
11. Click "get output".
12. This will bring up another page. Click "Send query to Galaxy".

This will retrieve 998 genes on chromosome 19 of mouse genome mm9. It is now necessary to convert this set into regions that correspond to the promoter regions. Here, we define the promoter region as a region 1000 bases upstream from a gene start site. Galaxy provides several tools to help users manipulate the genomic regions. Most of these tools are listed under "Text Manipulation" and "Operate on Genomic Intervals". Access them and convert your set as follows.

1. Click "Text Manipulation" to expand the list.
2. Select "Cut" operation.
3. Set "Cut columns:" to "c1,c2,c3,c4,c6", which correspond to "Chrom, Start, End, Name,Strand" in the input dataset.
4. Set "From:" to "15: UCSC Main on Mouse: refGene (chr19:1-61342430)".
5. Click "Execute".
6. Click on the "pencil" icon on the newly generated dataset "Cut on data 8".
   a. Set "Name:" to "RefSeq Genes (Interval)".

    b. Make sure that "Chrom column:" is set to "1".

    c. Make sure that "Start column:" is set to "2".

    d. Make sure that "End column:" is set to "3".

    e. Make sure that "Strand column" is checked and set to "5".

    f. Make sure that "Name/Identifier column" is checked and set to "4".

    g. Click "Save".

7. In the preview panel, make sure that the data format is "interval". If not, click on the "pencil" icon and select "Datatype". Set the "New Type:" to "interval".

8. Click "Operate on Genomic Intervals" to expand the list.

9. Select "Get flanks". This tool gets the flanking regions for a given dataset.

10. Set "Select data:" to "16: RefSeq Genes (Interval)".

11. Set "Location of the flanking regions/s:" to "Upstream".

12. Set "Length of the flanking regions(s):" to 1000.

13. Click "Execute".

14. Rename the dataset to "RefSeq Gene Upstream 1kb (Interval)".

15. Now, there are two datasets to be compared, the "RefSeq Gene Upstream 1kb (Interval)" and "CTCF Peaks, chr19, mm9 (BED)". To compare, select the "Join" operation under "Operate on Genomic Intervals" section.

16. Set "Join:" to "17: RefSeq Gene Upstream 1kb (Interval)".

17. Set "with:" to "9: CTCF Peaks, chr19, mm9 (BED)".

18. Set "Return:" to "Only records that are joined (INNER JOIN)".

19. Click "Execute".

20. Rename the dataset to "CTCF Peaks within 1kb upstream". There are 122 regions where the CTCF peaks overlap within 1kb upstream of known RefSeq genes on chromosome 19.

The same method can also be employed to investigate the overall distribution of the peaks in other genomic regions, such as intergenic or intronic regions. Other interesting analyses would be pathway analyses of the genes whose promoters overlap with the CTCF peaks. DAVID provides an extensive set of tools for annotating a list of genes and is well-suited for this purpose. DAVID can be accessed in the "Phenotype Association" section.

Another example of additional uses of this method is in the extraction of sequences for the binding sites and analysis with a motif discovery algorithm such as MEME. Galaxy provides a method for extracting genomic DNA sequences using the coordinates from the datasets. This operation can be done through the "Extract Genomic DNA" tool under "Fetch Sequences". Please note that this tool requires the dataset format be in a tabular format such as bed or interval formats, and that it be associated with a genome build. The extracted sequences can be saved to a local computer and analyzed with MEME algorithm. Unfortunately, the MEME algorithm is not available through Galaxy at this moment.

Fig. (6). Galaxy provides an easy method for retrieving data from the UCSC Table Browser. The History panel is hidden during this operation. You can click on the "<" icon at bottom right to display the History panel.

WORKFLOW

Previous exercises outline a few tasks usually performed during data analysis. While Galaxy provides a simple way to perform these tasks, repeating the same tasks over and over again upon multiple datasets can be cumbersome and error-prone. Galaxy allows users to create workflow to overcome this issue and allow scientists to focus on data interpretation.

Galaxy workflow provides a nice way to visualize the steps of an analysis and can be shared among researchers. This allows transparency and replicability in data analysis and also acts as a learning tool. To use or create the Workflow tool, it is necessary to have an account on the Galaxy server. To create an account, click on "User" in the top menu bar, select "Register" and fill in the details.

A workflow can be created either by constructing one from scratch or by extracting one from a working history. In the following exercise, we will build a workflow to predict CTCF binding sites using the mouse ChIP-Seq example above. We will use the same settings for each tool as given above.

Construct a workflow from scratch:

1. Log on to your Galaxy account.
2. Click "Workflow" in the menu bar. This will bring up the "Workflow" organizer page. If this is your first time to run the workflow tool, you will be presented with an empty list (Fig. **7a**).
3. Click "Create new workflow".
4. Set "Workflow Name:" to "MACS workflow", and click "Create".
5. A new workflow will be listed in the "Workflow" organizer page. At the current stage, the new workflow has 0 steps (Fig. **7b**).
6. Click on "MACS workflow" and select "Edit". This will bring up the "Workflow" editor. You can add a tool into your workflow from the set of tools in the Tool panel. Please note that not all tools can be included, for example, the "Upload File" tool is not available and must be included as an input dataset.
7. Scroll down to the bottom of the Tools panel under "Workflow control" and select "Inputs".
8. Click "Input dataset". This will create a box with the name "Input Dataset" in the editor. Name the box "Experimental Tags".
9. The next step is to groom the input dataset. Click on "NSG: QC and manipulation" to expand the list and select "FASTQ Groomer". By default, the workflow editor will place a tool in the middle of the screen. Arrange the "FASTQ Groomer" box to the right of the "Experimental Tags" box. Make sure the "FASTQ Groomer" uses the same settings as described in the example above.
10. Consolidate the previous two steps. Click on the ">" icon in the output row of "Experimental Tags" box to ">" icon beside "File to groom" of the "FASTQ Groomer" box as shown in Fig. (**8**). In this way, both steps are linked together by using the output from one tool as the input for the subsequent algorithm.
11. Click on empty space in the editor and drag the entire page to the left.
12. Click on "NGS: Mapping" to expand the list and select "Map with Bowtie for Illumina". Again, use the same settings as in the previous example.
13. Link the "output_file" from "FASTQ Groomer" box to the "FASTQ file" of "Map with Bowtie for Illumina" box.
14. Now, repeat steps 8 – 13 and label the input dataset "Control Tags".
15. The final step is to identify the binding sites. Click on "NGS: Peak Calling" to expand the list and select "MACS".
16. Link the "output (sam)" from the "Experimental Tags" steps to "ChIP-Seq Tag File" of "MACS" box and link the "output (sam)" from the "Control Tags" steps to "ChIP-Seq Control File".
17. Use the same settings for the MACS algorithm as those described in previous exercise.
18. Click on the star icon beside "output_bed_file (bed)" of the "MACS" box to mark it

as the output file of the current workflow. All non-flagged output files are hidden.

19. Fig. (9) shows the completed workflow. From the "setting" button on the top right corner of editor, you can save, run, and edit the workflow's attributes, re-configure the layout, and close the workflow.

20. Save the workflow and close it. You will be brought back to the organizer page. The "MACS workflow" is listed there in seven steps (Fig. 7c).

21. It is necessary to have input datasets before the workflow can be run. As an illustration, we will use the same mouse ChIP-Seq datasets again. Click on the "Shared Data" in the top menu bar, and select "Data Libraries". Select "ChIP-Seq Mouse Example", which selects both datasets and click "Go" to import to current history.

22. Click on "Analyze Data" to confirm that the datasets are now included in the History panel.

23. Click on "Workflow" again, then click "MACS workflow" and select "Run". The workflow is loaded into the Operation panel and is ready for execution (Fig. 10).

24. Set "Experimental Tags" to "2: Mouse ChIP-Seq Example Experimental Data, chr19, mm9", and set "Control Tags" to "1: Mouse ChIP-Seq example Control Data, chr19, mm9".

25. Click "Run workflow". During the execution, a number of files will be added to the history. Once the execution is completed, only those marked as output will remain.

26. Click on the "7: MACS on data 6 and data 5 (peaks: bed)" to expand the preview window. Here, we have identified 720 regions as the manual steps.

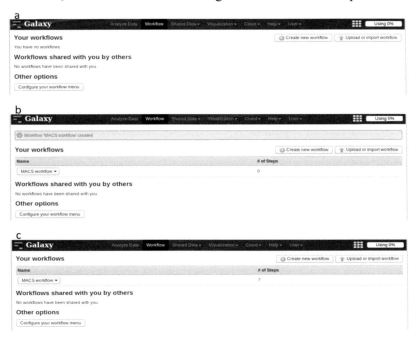

Fig. (7). The Workflow organizer shows an empty list **(a)**. You can create a new workflow from scratch or import an existing one. An empty workflow has been added to the Workflow organizer **(b)**. The newly created workflow has zero steps. The completed workflow has seven steps **(c)**.

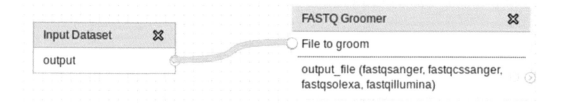

Fig. (8). To link two steps together, click on the ">" icon of an output to the ">" icon of an input field (File to groom). A line will be drawn between the two operations.

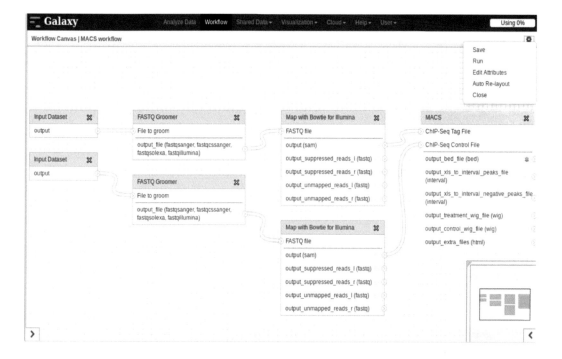

Fig. (9). Both the Tools and History are hidden in the expanded view of the workflow editor. You can show the Tool panel by clicking in the ">" icon on the bottom left corner. Click on the "<" in the bottom right corner to display the History panel. The steps have been re-organized using the Auto Re-layout option from the setting menu.

CONSTRUCT A WORKFLOW FROM A HISTORY

In the following example, we show how to create a workflow from a saved history. This method provides a quick way to construct a workflow from an existing pipeline. We will use the previous saved history from the Mouse ChIP-Seq example to build a workflow to find binding sites.

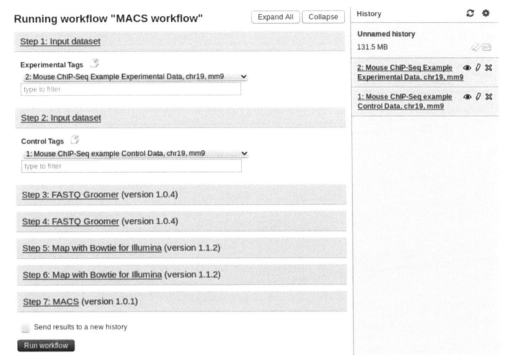

Fig. (10). The MACS workflow. The workflow will accept two input datasets, groom them, map the sequences to the mouse genome, and run the MACS algorithm on the mappings to predict the potential binding sites. Click on other steps to view the parameters used.

1. Click on the "setting" icon of the History panel, select "Saved Histories".
2. Click on the "Mouse ChIP-Seq". The datasets in the previous history will be loaded.
3. Click on the "setting" icon of the History panel again, and then select "Extract Workflow".
4. All previous working steps are listed in the middle panel (Fig. **11**).
5. Set "Workflow name" to "MACS Mouse ChIP-Seq Example Workflow".
6. As the "Upload File" tool is not available in the workflow, it is greyed out and replaced with "Treat as input dataset".
7. Since we want to create a workflow to identify the binding sites using MACS, uncheck the following tools,
    a. FastQC: Read QC
    b. UCSC Main
    c. Cut
    d. Get flanks
    e. Join
8. Click on "Create Workflow".
9. A new workflow is created. Click on "Workflow" in the top menu bar. The organizer shows two workflows, one built from scratch and the other extracted from an existing history.

10. Click on "MACS Mouse ChIP-Seq Example Workflow", and select "View". This will show a detailed description of the workflow, together with all settings.
11. Click on "Workflow" again, and edit the "MACS Mouse ChIP-Seq Example Workflow".
12. Flag the output_bed_ file of "MACS" algorithm to indicate that this file should be added to current history.
13. Save the workflow.
14. You can now run the new workflow on the example mouse datasets to validate the pipeline.

Fig. (11). Construction of a workflow by extracting the necessary steps from a history. Some tools, such as Upload File, cannot be used and must be included as an input dataset. Uncheck all the tools that will not be used.

REFERENCE

Trapnell, C., Roberts, A., Goff, L., Pertea, G., Kim, D., Kelley, D.R., Pimentel, H., Salzberg, S.L., Rinn, J.L., Pachter, L. (2012). Differential gene and transcript expression analysis of RNA-seq experiments with TopHat and Cufflinks. *Nature protocols, 7*, 562-578.

SUBJECT INDEX

Made in the USA
Columbia, SC
10 February 2020